AUTO BLUES

AUTO BLUES

A PRESCRIPTION FOR THE U.S. AUTOMOTIVE INDUSTRY

KAZUO OHMURA

PRENTICE HALL
New York London Toronto Sydney Tokyo Singapore

First published 1995 by
Prentice Hall
Simon & Schuster (Asia) Pte Ltd
Alexandra Distripark
Block 4, #04-31
Pasir Panjang Road
Singapore 0511

Printed in Singapore

1 2 3 4 5 99 98 97 96 95

ISBN 0-13-109240-5

Prentice Hall International (UK) Limited, *London*
Prentice Hall of Australia Pty. Limited, *Sydney*
Prentice Hall Canada Inc., *Toronto*
Prentice Hall Hispanoamericana, S.A., *Mexico*
Prentice Hall of India Private Limited, *New Delhi*
Prentice Hall of Japan, Inc., *Tokyo*
Editora Prentice Hall do Brasil, Ltda., *Rio de Janeiro*
Prentice Hall, Inc., *Englewood Cliffs, New Jersey*

Contents

List of Tables vii

List of Figures viii

Preface ix

Introduction xiii

■ Part I The American Automotive Industry in the Doldrums

1 That Sinking Feeling 3
2 Riddles 9
3 An Overview of Past Analyses 18

■ Part II From Heaven to Hell

4 Stable Demand, Deteriorating Output 29
5 The Waning Appeal of the New Car 36
6 Desperate Efforts at Innovation 46

■ Part III Deviating from the Average

7 Immobile Consumers 59
8 Limited Availability 69
9 Growing Income Dispersion 78

■ Part IV A Vicious Circle

10 Target Marketing 87
11 Rich Man, Poor Man 93
12 Management Problems 100

■ Part V Breaking through into the 1990s

13 A Flood of Jalopies: The Hidden Potential 107
14 Advanced Standard Strategy 112
15 Direction of the Big Three 117
16 Japanese Manufacturers' Options 124

■ Part VI Lessons from the U.S. Automotive Industry

17 Status of the World Automotive Industry 131
18 Dynamics of the World Automotive Industry 144
19 Future Cars: A Concept 153

Bibliography 160

Index 165

Tables

Table 1.1 Operating results of the Big Three, 1980–92
Table 2.1 World automotive industry production and sales, 1992
Table 2.2 Top twelve global vehicle producers, 1992
Table 2.3 U.S. supply and demand of vehicles
Table 2.4 Japanese transplant production
Table 2.5 Big Three market share of passenger cars
Table 2.6 U.S. vehicle trade, 1991
Table 3.1 Wage rate comparison: Hourly earnings (in dollars) of production workers
Table 3.2 Motor vehicle and equipment manufacturing production workers' hourly compensation in selected countries, 1988
Table 4.1 Demand, output, and net import of automobile sector
Table 5.1 Personal consumption expenditures for transportation as percentage of total personal consumption
Table 5.2 U.S. retail sales of new cars by sector
Table 5.3 Index of passenger car operating costs
Table 5.4 Retail car sales by class
Table 5.5 Segment-by-segment market share in 1991
Table 8.1 Consumer expenditure per new car, new car prices, and annual median income
Table 8.2 Trend of lowest new car price
Table 8.3 Car price comparison: Domestic vs. import
Table 9.1 Family income at selected positions
Table 9.2 Percentage of aggregate income of families
Table 10.1 Auto price, top earner income, and industry wage rate
Table 10.2 Outline of auto-related industry
Table 15.1 A strategic matrix of automotive manufacturers
Table 17.1 Global vehicle demand

Figures

Figure 2.1	Passenger cars: Sales vs. cars in use
Figure 2.2	Average age of vehicles in use
Figure 4.1	Vehicle output as percentage of GNP
Figure 4.2	Automotive demand and output as percentage of GNP
Figure 5.1	Personal consumption: New cars, used cars, and other vehicles
Figure 6.1	Capital spending per vehicle produced
Figure 6.2	Production capacity
Figure 6.3	Labor productivity: Vehicles produced per employee
Figure 7.1	Monthly market share of the Big Three and Japanese cars
Figure 9.1	Gini ratios of family income
Figure 9.2	Distribution of families by income class
Figure 9.3	Share of net worth of American families
Figure 10.1	Three indices: Average car price, top 5 percent income levels, and autoworker wage rates
Figure 13.1	Cars in use by model year
Figure 13.2	Average fuel economy: New car standard vs. overall average
Figure 17.1	Average shipment price index: Passenger cars vs. trucks
Figure 18.1	Per capita GNP/GDP and vehicle diffusion rate
Figure 18.2	Economic development and the automotive industry: Normal case
Figure 18.3	Economic development and the automotive industry: Intermittent technology transfer case

Preface

This book is divided into six parts.

Part I focuses on the current state of the American automotive industry. It also describes how other researchers have examined the problems in the industry. Readers already familiar with the industry may wish to skip this part.

Part II offers a bird's-eye view of the automotive industry. It depicts the industry in terms of the supply-and-demand situation using macro-economic data that are not normally applied to the automotive industry. While the findings themselves are not new, this part summarizes the economic characteristics of the industry.

Part III examines the characteristics of the 1990–91 recession. (Specialists on the automotive industry may choose to begin here.) The approach taken here is similar to that employed in the marketing of automobiles. It scrutinizes the demographics to determine what is happening in the marketplace.

Part III also offers an insight into how Japanese manufacturers analyze the American market. This will provide market researchers at the three major U.S. automakers, or the Big Three, with an opportunity to view the thinking pattern of Japanese auto manufacturers. The author knows from experience that the Japanese automakers have accumulated data on the American market that rivals that of their American counterparts. An examination of the market outlook is essential to developing new models and drawing up marketing strategies. The Japanese success has been achieved, in part, due to their unparalleled ability to analyze the American market. Part III will offer information that may be of use to American automakers.

Part IV combines an economic analysis of the industry with an examination of the strategies that have been pursued by the automotive companies. More specifically, this part tries to identify the strategies that the automakers have pursued to increase their profits against the backdrop of slower economic growth, widening income dispersion, and the rigid manner in which wage rates are determined in the United States. It will show that the Big Three did indeed try to draw up the optimum strategy, but that their competitors were also adopting similar strategies. Consequently, the Big Three face the problematic situation that we see today.

Part V suggests some options for the future based on the findings set forth in previous parts of this book. The options available to the government, the Big Three, and the Japanese manufacturers are examined. Part V sets forth the thesis that the American market is not in a state of ruin yet; rather, it is building up potential demand. The issue here is whether this potential can be realized. The time is ripe for the Big Three to be restored.

Part VI describes the implications the above study of the U.S. automotive industry presents to world automotive companies. A slow-down in vehicle demand is observable commonly in the most advanced countries, such as the United Kingdom, Germany, and Japan as well as in the United States. To encourage global automotive demand, a necessary condition is to provide developing countries with cheap vehicles. As manufacturing newcomers from such countries are limited because of the gap in the technology needed to clear many kinds of regulations, the role of existing automotive companies should be greatly enhanced. They must globalize their operations and manufacture vehicles with appropriate features. The concept of future cars also will be discussed in this part.

Readers are advised to keep two basic points in mind. The first point concerns the distinction between "new cars" and "used cars." Though demand for automobiles tends to be considered in terms of "demand for new cars," this type of demand is determined annually by the general demand for automobiles, which is inclusive of used cars. In other words, cars "in use" is the best indicator of the basic demand for automobiles. New cars will sell if manufacturers succeed in offering cars that are more attractive than the cars that are in use, or when over-aged cars are scrapped and replaced by new ones. The reverse is also true. The earnings of automotive manufacturers are determined by how many new cars are sold, so the focus tends to be placed there; the significant thing to look out for, however, is the demand for cars in general.

The second point to be kept in mind concerns the word "median," which is used frequently in conjunction with the discussion on income dispersion. The median differs from the arithmetical average in the following manner.

Take the example of nine households in which one household has an annual income of $100,000, and the rest, $10,000. (All dollars are in U.S. currency unless otherwise stated.) If the arithmetical average of this group ($20,000) is deemed to be the average for these households, we face two problems: (1) not a single household actually earns $20,000, and (2) the minority (in this case, one household) excessively skews the final outcome. The median, on the other hand, takes the income earned by the fifth ranking household ($10,000 in this case) as the "average" for the group. Thus, the median serves as a better indicator than the arithmetical average. The disadvantage is that the median does not tell us the total income for the group, which is $180,000 in this case. However, the median concept is used frequently in statistics on income distribution.

In closing, I would like to note that I was showered with help and support in writing this book. It would not have been possible for an outsider like myself to complete a book of this sort without help from those patient enough to teach me the A to Z of the industry. In that sense, I must thank all the people that I have met during the past decade or so.

I would like to express my gratitude especially to Mr. Takashi Uzawa, Manager of the International Public Affairs Division of Toyota Motor Corporation, and Mr. Daizo Kuyama, Manager of the Product Management Division of Toyota Motor Corporation, who have been kind enough to read through the book as well as to advise me on certain points. I would also like to thank Mr. Ryuzo Murayama, Associate Professor of the Osaka University of Foreign Studies, and Mr. Naoya Tsutiya, Staff Writer of the Economic News Department of Editorial Bureau of Nihon Keizai Shimbun, who gave me valuable comments and suggestions. Special thanks go to Mr. Scott Merlis, Executive Director of Morgan Stanley Inc., who covers the U.S. automotive industry in New York, and Mr. Michael Chinworth, Senior Analyst of the Asian Technology of The Analytic Sciences Corporation, who carefully read through the manuscript and gave me very helpful comments. I hasten to add, however, that any errors in this book are attributable to myself alone.

I wish to express my special thanks to Juri Yamamoto, of Kinyu Keizai Honyaku Jimusho, who assisted me in translating the Japanese text into English; to Miwako Kiuchi, formerly with the research department of Morgan Stanley, for her help in gathering research material; and to my editor, William Auckerman. My greatest thanks, however, goes to my wife, Mineko, who worked hard to ensure that I worked in comfort.

Introduction

In the automotive industry, more so than in any other, the images projected by the manufactured products differ markedly from the perceived images of the industry itself. The typical car buyer is interested only in the features of the automobile: Is the design attractive, and the styling fantastic? Can the engine outperform other engines? Does the car have outstanding drivability, lots of loading room, great acceleration, and superb aerodynamics? Are the colors attractive? Is the interior impeccable, filled with advanced electronic components and all the newest features? Does the car project an "up-market" image?

Naturally, consumers are very conscious of who manufactures what, but they understandably take little or no interest in a particular company's actual management policies or production set-up. What attracts the attention of consumers are the carefully cultivated images of the various brands. Cadillac, Lincoln, Lexus, Mercedes, and Jaguar, for example, are perceived as being "luxury" cars. Chevrolet, Corolla, and Volkswagen, on the other hand, are considered to be "practical" cars.

Obviously, manufacturers do not deliberately pursue strategies that might be detrimental to their established brand images, but can the average consumer really tell whether an "image" conforms to "facts"? Can a potential buyer judge whether a so-called luxury car really outperforms a cheaper car, or whether, in the long run, a low-end model will be more cost-effective than a luxury model? Yet, if they are to remain economically viable, manufacturers must differentiate their various models on the basis of cost performance without making this effort too obvious to the average consumer.

What might the automotive industry look like if the manufacturers and consumers could be brought closer together? Let us consider Broadway theaters and theatergoers as an analogy. The manufacturers would be the theaters that line the streets in the theater district of New York; the consumers would be the pedestrians strolling through the theater district. There are large billboards throughout the area, and posters are pasted on the walls almost on top of each other, trying to lure an audience. Some of the people already know exactly which show they want to see. Others, walking by, might stop in and watch a show if they notice an advertisement for one that looks interesting. Still other people, though, have no intention of entering any of the theaters no matter how enticing the advertisements may be.

The factors behind the rise or fall of any industry are much like the situation described above. The industry cannot hope to prosper if people flock to just one theater and not to the others. Nor can it hope to prosper without appealing to a diverse potential audience. Industry-wide prosperity can be achieved only as a result of numerous triumphs, both big and small, by each individual theater. It is only through such individual triumphs that people will gather, from across the country and around the world, to see the shows. To ensure success—a continuous stream of people visiting the theaters—newspapers and magazines must review the shows, and audience members must later tell their family and friends.

This book looks at the automotive industry from three different perspectives. First, it takes a bird's-eye view of the automotive industry, much like looking down at the theater district from the Empire State Building. Rather than focusing on individual automakers, it tries to adopt a balanced viewpoint by looking at the issues both from the consumer's perspective and from the manufacturers' perspective.

Individual corporations cannot be completely ignored, however, because they play a major role in the rise or fall of an industry. What can save the declining fortunes of Broadway, for example, are the efforts made by the individual theaters in the face of such decline; they must each identify the problems and seek solutions.

Viewed from the individual theater's perspective, it is entrepreneurship that creates shows. Entrepreneurs might respond to declining audiences by opening innovative theaters to provide the public with shows that open at different hours from traditional performance times, or they might offer lower-priced tickets. We cannot know what course an industry could take unless we examine the methods by which corporations can overcome their current difficulties. The industry will decline only if its participants lose interest and stop making efforts to save it.

When we examine the individual theaters more closely, however, we find that what the audience sees and what the operators see are very different. The audience's interest is focused on the show itself: the performers, the plot, the effects, the props—in other words, only on what is visible onstage. The operators, however, must concentrate on totally different concerns: negotiating with the scriptwriters and directors, casting the roles, procuring the necessary props, determining the ticket prices, advertising, and much more. And once a show is underway, they must begin planning for the next show. The operators cannot sit back and enjoy the show with the audience.

Obviously, the theaters must present shows that appeal to potential theatergoers; otherwise, they will not be able to attract an audience no matter how many people might stroll past. The situation faced by the U.S. automotive industry is similar; while the number of "strollers" has not declined, the "attendance" is much lower. If this trend continues, people may come to prefer attending shows in London, Berlin, Paris, or Tokyo. Just because the number of people who like to go to the theater does not decline is no guarantee that the local theaters will continue to thrive.

It is too simplistic to point a finger at the industry leadership and lay all the blame on them. While there are obvious problems associated with management, we must identify the root sources of such problems. Are the problems due to the peculiar personality of the individuals concerned, to the corporate culture (something ingrained in the industrial organization), or to something else entirely?

Thus, this book also looks at the automotive industry from the individual firms' perspective: it examines some of the micro issues that the leadership of the Big Three may have overlooked. Returning once again to the Broadway theater analogy, one way to drum up an audience when business is slow is to put on a show that appeals to a select core of loyal fans. Rather than trying to attract the general passersby (that is, all potential customers), the theater operators instead target specific customers who are more prone to spend their money on the theater.

Unfortunately, as this book will show, this "brilliant" targeting strategy ultimately leads to long-term decline. This is not to say that the strategy *never* works; it may work at one time or under certain circumstances, yet be totally inappropriate for other times. The leadership of automotive companies have failed to recognize the limitations of this strategy. It is the sort of strategy that could only have been conjured up in a market that automakers felt they knew all too well. It is for this very reason that they have failed to recognize their mistake.

Third, the scope of our inquiries and discussions is not limited to the American marketplace, nor to the activities of U.S. automakers. In many cases, no distinction is made between domestic and foreign vehicles. While the three major U.S. automakers are the main focus of this book, it is impossible to ignore the Japanese and other foreign manufacturers who are also major suppliers of cars to the American market. It is necessary that we examine whether the penetration of imported cars contributes to the overall prosperity of the industry, or whether foreign manufacturers are instead cannibalizing the market developed by domestic automakers.

If the foreign manufacturers have developed their own new markets, then the Big Three need not consider them as direct threats. They may even have contributed to increasing the overall size of the market. If foreign manufacturers are in fact taking customers away from the Big Three, however, and the overall market size has not grown, then the Big Three will face difficult times ahead. Determining the true impact that imported cars have had on the domestic industry is thus significant both for the U.S. government (which must develop appropriate administrative measures) and for foreign and domestic automakers (which need to establish appropriate corporate strategies).

PART I

The American Automotive Industry in the Doldrums

1

That Sinking Feeling

Bush to Visit Tokyo with U.S. Auto Execs

When Japanese automotive executives saw this newspaper headline in December 1991, their initial reaction was disbelief: U.S. President Bush coming to Tokyo with the leadership of the Big Three? Their disbelief soon gave way to consternation as the troubling implications emerged. What new demands did the Americans intend to make of the Japanese automotive industry? Would the American president now lend his prestige to their continuing gripes about Japanese foreign trade policy?

It soon became apparent to all that the situation facing the U.S. automotive industry was grim. Another newspaper headline less than two weeks later could only fill the Japanese automotive industry with foreboding, indicating the probable tenor of the upcoming U.S.-Japan talks:

GM to Close 21 More Plants, Cut Outlays and 74,000 Jobs[1]

The accompanying article revealed that, in a major cost-cutting effort, General Motors (GM) would close down six auto assembly plants, four auto engine and transmission plants, and eleven auto parts manufacturing plants. Not only did GM intend to suspend hiring during 1992, but some 74,000 blue-collar workers would lose their jobs, and another 9,000 white-collar jobs would be slashed. Moreover, assets unrelated to the auto business were to be disinvested, and the cost reduction program would continue on into 1995.

These drastic measures apparently came too late, though—GM's strength continued to dwindle. Some analysts even began to ask the unthinkable: Could the world's largest auto manufacturer continue to sustain itself? As an article in *Fortune* magazine put it: "CAN GM REMODEL ITSELF? The company can make great cars. But can it make money on them? As losses pile up and cash dwindles, CEO Stempel orders major cuts but needs to do more."[2]

GM was already finding it difficult to maintain its number one position in the global automotive industry. Although it remained on top in terms of global production in 1991 (ahead of Ford and Toyota), GM had slipped to number two (behind Toyota) in terms of domestic production.[3]

February 1992 brought still more disturbing news. Each of the Big Three automakers announced large operations deficits in 1991. GM's 1991 deficit was a record-high $4.45 billion, which came hard on the heels of its previous record-high loss of $1.98 billion in 1990 (see Table 1.1).

The source of GM's record deficit was its loss on sales of cars within the American market. GM's domestic sales showed a massive loss of $7.09 billion (following a loss of $4.57 billion in 1990). This was only partially offset by GM's $1.76 billion profit in Europe (where it managed to expand its market share in Germany) and modest profits from sales in Mexico and other Central and South American markets. In addition, GM's related businesses—Hughes Aircraft, Electronic Data Services (EDS), and General Motors Acceptance Corporation (GMAC), its financing subsidiary—all posted substantial profits. These profitable operations, however, could not offset the massive red ink figure of U.S. auto sales.

Nor was GM the only U.S. automotive manufacturer to face a financial dilemma. The U.S. recession also affected the domestic car sales of Ford, which posted an after-tax loss of $2.26 billion in 1991 (a substantial deterioration from its $860 million profit in 1990). Ford of Britain and Jaguar also posted losses of $761 million and $354 million, respectively. Ford business in the markets outside of the U.S. accounted for $970 million of Ford's overall loss.

Chrysler, which does not possess a large production base outside of the U.S., posted an overall after-tax loss of $810 million. Chrysler suffered over $1.16 billion of losses in the U.S., but it

Table 1.1 Operating results of the Big Three, 1980–92 (million dollars)

Year	General Motors		Ford Motors		Chrysler		Big Three total	
	Sales and revenues	Net income	Sales	Net income	Sales	Net income	Sales and revenues	Net income
1980	62,698	(775)	37,085	(1,543)	8,600	(1,710)	108,383	(4,028)
1981	69,191	320	38,247	(1,060)	9,972	(476)	117,410	(1,216)
1982	67,747	962	37,067	(657)	10,057	170	114,871	475
1983	82,502	3,730	44,454	1,866	13,264	701	140,220	6,297
1984	93,144	4,516	52,526	2,906	19,573	2,373	165,243	9,795
1985	106,655	3,999	52,915	2,515	21,025	1,610	180,595	8,124
1986	115,609	2,944	62,868	3,285	21,937	1,389	200,414	7,618
1987	114,870	3,550	71,797	4,625	25,381	1,290	212,048	9,465
1988	123,641	4,856	82,193	5,300	30,790	1,050	236,624	11,206
1989	126,931	4,224	82,879	3,835	31,039	359	240,849	8,418
1990	124,705	(1,985)	81,844	860	26,965	68	233,514	(1,057)
1991	123,056	(4,452)	72,050	(2,258)	26,707	(795)	221,813	(7,505)
1992	132,429	(23,498)	84,407	(7,385)	33,409	723	250,245	(30,160)

( ) = deficits

Note: In 1992, a one-time charge to net income, due to a change in the accounting standard for retiree health care, was appropriated by GM and Ford which amounted to $20.8 billion and $7.5 billion, respectively.

Sources: Annual reports.

managed to post fourth-quarter 1991 profits that included $205 million gains on equity investment transactions.

Lee Iaccoca, chairman of Chrysler, was optimistic, however, especially about the new interest in leisure vehicles cultivated by Chrysler's minivans.

> [O]ur operation results for the fourth quarter reflect the impact of our $3 billion cost-cutting program. We have had to manage our way through the worst market since 1983.... More importantly, for 1991, we increased our North American combined car and truck market share to 12.4 percent while our major domestic competitors lost ground. Our minivans had their second best sales year ever and continued to dominate their market segment.[4]

The background against which the U.S.-Japan summit talks opened on January 7, 1992, then, was this: the U.S. automotive industry, which used to signify the American way of life, faced dwindling sales of passenger cars. As the Japanese and American automotive leaders met in conjunction with the governmental meetings, the expressions on their faces (seen during internationally televised interviews) told the story. The American leaders made frank comments about the progress of the talks, but the Japanese executives held back and kept quiet. It was the usual reaction on both sides—the Americans taking the offense and the Japanese on the defense.

Not unexpectedly, the issue on which the American CEOs chose to focus was the Japan-U.S. trade imbalance. In 1991, the U.S. had imported $21.2 billion worth of Japanese cars, while Japan imported only $580 million worth of cars from the U.S.[5] Overall, American cars accounted for less than 1 percent of total car sales in Japan, while Japanese cars made up 30 percent of total car sales in the U.S.[6]

A closed auto distribution network, close ties between the auto and parts manufacturers, and implicit governmental guidance suggest that the Japanese auto market is not as open as the American market. The viewpoint of the Japanese industry, however, is that the difference in the number of imports is a consequence of the ability to compete in the international market—period. They note that Japan does not levy an import duty on imported cars (the only country in the world that does not do so).

The Japanese automotive companies further assert that today's success is built upon the series of hardships that they experienced when they first tried to enter the American market. In turn, they ask, "How hard have the Americans tried to seriously penetrate the Japanese market?" and, "If the Japanese market is so closed, how do you explain the popularity of the German cars in this market?" (In fact, German imports outnumber American Big Three imports in Japan by nearly 9 to 1.[7])

Although the automotive summit meeting failed to close the gap between the two parties, an import target was agreed upon. The five top Japanese auto makers (Toyota, Nissan, Honda, Mitsubishi, and Mazda) agreed to import 19,700 American cars—a drop in the bucket for the world's second largest market! The target date for achieving this is 1994 and 1995. The agreement was, nevertheless, seen as a beginning by the American automotive leaders.

The January meeting also resulted in an automotive parts import agreement. Japanese automakers agreed to increase the amount of imported auto parts into Japan (from the $7 billion recorded in 1990) to $15 billion in 1994. At the same time, the required local content ratio of Japanese transplants in the U.S. will be raised from 50 percent to 70 percent.

In short, the outcome of the first meeting was less than startling. The American leaders were promised a minuscule increase in the number of American-made car imports into Japan. The Japanese, on the other hand, were left with a parts import target that they will find virtually impossible to achieve. The net result can only be described as "both sides lose." The problems of the Big Three were not resolved, yet the agreement will be a burden on the Japanese automakers for many years to come.

A second meeting of the top managers of the auto industry was held on May 26, 1992. This meeting merely provided a forum to confirm the agreements reached previously. The significance of these two meetings was that the industry leadership of both countries finally met face-to-face to discuss the issues confronting them. Neither side had really expected the complex issues to be resolved during the course of the meetings. As one participant put it, "The important thing is that we participated!"[8]

It is noteworthy that, following the series of grim corporate earnings announcements and the first Japan-U.S. automotive summit meeting, a number of leadership changes took place in Detroit. On March 16, 1992, Lee Iaccoca appointed Robert Eaton of GM as his successor. Eaton joined Chrysler after having spent four years as the CEO (Chief Executive Officer) of General Motors Europe. And on April 4, 1992, a major reshuffle took place within the GM hierarchy primarily because the necessary cost-cutting programs were not being implemented on time. The position of president was handed over from Lloyd Reuss to Jack Smith, and that of CFO (Chief Financial Officer) from Bob O'Connell to William Hoglund. The leadership was thus passed on from those who had been responsible for operations in North America to those responsible for overseeing operations in Europe. The power structure thus shifted toward outside directors, something new for GM.

The operating results for the first quarter of 1992 (announced in late April) surpassed most expectations. Although Chrysler posted losses, both GM and Ford turned around to post profits. The improved earnings were the result of smaller discounts granted in the U.S. market and cost reduction measures. GM, however, continued to suffer a loss ($1.1 billion) from its U.S. operation.

The steady improvement in demand, supported by demand for small trucks, became even more evident by the April–June quarter. The 1991 recession seemed to be on its way out. Yet hardly a soul thought that the problems of the U.S. automotive industry were behind them.

Plans to enhance competitive ability and programs to reduce costs have been repeatedly implemented; it is too optimistic, however, to assume that the measures taken thus far can offer fundamental solutions to the major problems that exist. Even the problems themselves have not yet been clearly defined.

The issues are mounting—this is clear to all. The automotive industry is suffering, and the recent improvement in earnings may provide the last opportunity to revamp the American automotive industry. But what else needs to be done to resuscitate the U.S. automotive industry? This decade-old question has yet to be answered.

Notes

1. *Asian Wall Street Journal*, December 19, 1991.
2. Alex Taylor III, "Can GM remodel itself?" *Fortune*, January 13, 1992, pp. 20–26.
3. In 1991, GM's global production was just over 7 million units, well ahead of Ford (5.4 million) and Toyota (4.7 million). GM produced only 3.6 million cars in the U.S., however, while Toyota produced 4.1 million cars in Japan (*Market Data Book*, Detroit: Automotive News, 1992, pp. 3, 10, 14).
4. *PR Newswire*, February 6, 1992.
5. Figures are on a 1991 FAS (free alongside ship) customs basis (*Automotive News*, April 6, 1992).
6. The total number of vehicles sold in Japan during 1991 was 4,028,000 (excluding subcompacts). The 30,128 vehicles imported from the U.S. accounted for a mere 0.7 percent of that total (and only 15.3 percent of Japan's total vehicle imports). Nearly half of the U.S. exports to Japan (14,302 vehicles), however, were "made-in-America" Hondas. Big Three exports to Japan in 1991 were GM, 9,261 vehicles; Ford, 2,959; and Chrysler, 1,491.
7. Japan Tariff Association, *Japan Exports and Imports, Commodity by Country*, December 1991. Exports are on an FOB (free on board) basis, while imports are on a CIF (cost, insurance, and freight) basis. Auto trade between Japan and Germany is relatively balanced on a monetary basis, if not a unit basis. In 1991, Japan imported $3.4 billion of vehicles from Germany (115,000 cars) while exporting $4.8 billion (569,000 cars).
8. "*Ometemuki 'Seika'—Nichi-Bei Jidosha Samitto* (Ostensibly 'fruitful'—Japan-U.S. automotive summit)," *Asahi Shimbun*, May 20, 1992.

2

Riddles

How big is the U.S. automotive industry? Is the demand for passenger cars in the U.S. increasing or declining? Though at first glance these might appear to be simple questions, the answers are exceedingly complex.

Automotive statistics are considered a leading national economic indicator. Of the twelve indicators relating to personal consumption traditionally used to estimate the U.S. GNP, four are auto-related.[1] Data pertaining to the number of cars produced and sold, as well as to the prices of cars, is systematically organized, so gathering statistics in this area is not a laborious process. Interpreting those statistics, however, is a difficult and complicated endeavor.

Table 2.1 provides a breakdown of worldwide vehicle production and sales in 1992 by country; Table 2.2 lists the top twelve global vehicle manufacturers. As Table 2.1 shows, the U.S. holds a clear lead in the number of cars purchased, and it ranks second (after Japan) in the number of cars produced.[2] For total vehicle production, the record year for the U.S. was 1978, when the American automotive industry produced 12.89 million units. If we consider only passenger car production, however, the peak year was fully twenty years ago, in 1973, when 9.66 million passenger cars were produced.

Table 2.1 World automotive industry production and sales, 1992 (thousand units)

	Vehicle production	Vehicle sales
U.S.	9,777.9	12,885.7
(U.S. share of total)	(20.6%)	(28.3%)
Canada	1,982.9	1,210.1
U.K.	1,540.3	1,794.8
Germany*	5,194.0	3,629.7
France	3,763.2	2,428.0
Italy	1,686.1	2,524.5
Sweden	302.6	166.1
Spain	2,303.7	1,283.3
Mexico	1,083.1	706.9
Brazil	1,091.8	756.8
Japan	12,499.3	6,959.1
S. Korea	1,729.7	1,724.5
Australia	209.4	356.6
Others	4,277.6	9,090.1
World total	47,441.6	45,516.2

*Including East Germany

Source: *Market Data Book*, Automotive News, 1993.

Table 2.2 Top twelve global vehicle producers, 1992 (thousand units)

GM	7,146
Ford	5,764
Toyota	4,695
Volkswagen	3,499
Nissan	2,982
Fiat	2,231
Chrysler	2,159
Peugeot-Citroen	2,050
Renault	2,042
Mitsubishi	1,832
Honda	1,828
Mazda	1,460
Big Three total	15,069
Big Three share of world vehicle production	31.8%

Source: *Market Data Book*, Automotive News, 1993.

The decline has continued for two decades, and the structure of the U.S. automotive industry has become more complicated. "Made-in-America" no longer means just cars produced by the Big Three; the output of Japanese transplants in the U.S. has increased dramatically. And "imports" is no longer synonymous with "imported cars." The Big Three themselves market captive imports from Japan and Korea alongside those from Canada and Mexico. Even Japanese cars are being imported from Canada.[3] The distinction is becoming further blurred, with a growing proportion of imported parts being used in cars assembled domestically by the Big Three. In some cases, the foreign-content ratio is even higher than that of their Japanese counterparts.

Combined passenger car and truck sales in the U.S. ranged between 10 and 12 million units during the recessionary years of 1982 and 1992, far lower than the 15 to 16 million units sold annually during the boom years of 1978 and 1985. Sales of passenger cars have remained between 8 and 11.5 million units for the past twenty years. The average number of cars sold annually during the 1980s (9.8 million) was lower than that of the 1970s (10.2 million). Furthermore, the proportion of new "American made" cars has been steadily declining.

Table 2.3 shows the annual supply and demand for vehicles in the U.S. market from 1980 to 1992. The proportion of imported vehicles (passenger cars and trucks) reached a peak of 26.7 percent in 1987, then declined to 17.8 percent in 1992.[4] This decrease, however, is largely a result of increased production by Japanese transplants.

As Table 2.4 shows, the number of vehicles produced by these transplants has increased at a rapid pace. In 1992, transplants accounted for 17.3 percent of all vehicles produced in the U.S. The 1.69 million vehicles manufactured by transplants in 1992 was more than the 1.28 million manufactured within the U.S. by Chrysler. Honda's Marysville, Ohio, plant is the largest single passenger car assembly plant in the U.S., turning out nearly a half-million vehicles annually. In only a decade, Japanese transplants have made the U.S. a major production base.

Let us turn now to the penetration of Japanese vehicles into the American market. In this context, the term "Japanese vehicles" should be understood as including those produced by the transplants. As Table 2.5 shows, the market share held by the Big Three (including their captive imports) declined during the 1980s.

The market share for foreign vehicles briefly remained flat after voluntary export quotas were set by the Japanese

Table 2.3 U.S. supply and demand of vehicles (thousand units)

Year	Domestic production	Retail sales			Import	
		Cars	Trucks	Total*	Import	Ratio (%)
1980	8,009	8,979	2,487	11,464	2,884	25.2
1981	7,942	8,536	2,260	10,796	2,778	25.7
1982	6,985	7,982	2,560	10,542	2,637	25.0
1983	9,224	9,182	3,129	12,312	2,858	23.2
1984	10,924	10,390	4,093	14,484	3,057	21.1
1985	11,652	11,042	4,682	15,724	3,617	23.0
1986	11,334	11,460	4,863	16,322	4,186	25.6
1987	10,924	10,277	4,912	15,189	4,053	26.7
1988	11,213	10,530	5,149	15,679	3,645	23.2
1989	10,874	9,772	4,941	14,713	3,237	22.0
1990	9,782	9,300	4,846	14,147	3,036	21.5
1991	8,810	8,175	4,365	12,539	2,589	20.6
1992	9,703	8,214	4,904	13,118	2,329	17.8

*Due to rounding errors, total retail sales data do not always coincide with the sum of the retail sales of cars and trucks.

Sources: 1980–1991 data: MVMA, *Facts & Figures*, 1992; 1992 data: *Ward's Automotive Reports*, January 18, 1993.

Table 2.4 Japanese transplant production (thousand units)

Year	Cars	Trucks	Total*	% U.S. production
1982	1.5	–	1.5	–
1983	55.3	19.9	75.2	0.8
1984	138.5	100.5	239.0	2.2
1985	253.7	107.4	361.2	3.1
1986	509.1	108.0	617.1	5.6
1987	633.0	102.7	735.7	6.7
1988	794.4	95.9	890.3	7.9
1989	1,130.9	123.1	1,253.9	11.5
1990	1,310.0	173.9	1,483.9	15.2
1991	1,356.2	192.4	1,548.7	17.6
1992	1,416.4	271.1	1,687.4	17.3

*Total production of vehicles produced by Diamond-Star, Honda, NUMMI, Nissan, Toyota, Auto Alliance and Subaru-Isuzu.

Sources: 1982–91 data: *Ward's Automotive Yearbook*, 1988 and 1992; 1992 data: *Ward's Automotive Reports*, January 18, 1993.

Table 2.5 Big Three market share of passenger cars (thousand units)

Year	Big Three*	% share	Imports	% share	Total
1980	6,624.9	73.8	2,351.2	26.2	8,976.1
1981	6,218.7	73.0	2,295.9	27.0	8,514.6
1982	5,805.4	73.0	2,151.0	27.0	7,956.4
1983	6,802.7	74.4	2,345.9	25.6	9,148.6
1984	7,861.1	76.1	2,462.8	23.9	10,323.9
1985	8,140.2	74.1	2,838.6	25.9	10,978.8
1986	8,155.8	71.5	3,248.3	28.5	11,404.1
1987	6,885.5	67.6	3,301.9	32.4	10,187.4
1988	7,303.0	69.3	3,240.0	30.7	10,543.0
1989	6,634.6	67.9	3,142.8	32.1	9,777.4
1990	6,105.3	65.6	3,194.9	34.4	9,300.2
1991	5,248.0	64.2	2,926.6	35.8	8,174.6
1992	5,301.1	64.5	2,913.0	35.5	8,214.1

*Before 1986, American Motors' sales units are included. Big Three includes captive imports.

Sources: *Ward's Automotive Yearbook*, 1992 and 1988; *Ward's Automotive Reports*, January 18, 1993.

manufacturers in 1984/85, but it began to increase again shortly thereafter. As Table 2.5 shows, the total number of vehicles sold during 1990/91 was below that of 1988, but the market share held by foreign vehicles increased to a record 35.8 percent.[5] Even though various forms of import restrictions have been levied by the American government on the importation of Japanese vehicles, their market share continues to rise. In 1989, the Honda Accord became the best selling car in the U.S., surpassing the popularity of the Ford Taurus.

The U.S. is the largest consumer of automobiles, but Japan is the leading producer. The U.S. automotive industry now manufactures only 69 percent of all vehicles sold in the U.S. In other words, the U.S. market is heavily reliant on imported vehicles. On the other hand, the U.S. exports significant numbers of vehicles only to Canada. Other important markets are the Middle East ($1.6 billion in 1991), Taiwan ($612 million), and Central/South America ($576 million). The overall result is a substantial deficit in the U.S. automobile trade, as shown in Table 2.6.

The U.S. posted a total trade deficit of $66.2 billion in 1991;

Table 2.6 U.S. vehicle trade, 1991 (million dollars)

Country	Import	Export	Balance
Australia	242.1	11.2	(230.9)
Belgium	296.3	270.9	(25.4)
Brazil	70.2	38.4	(31.8)
Canada	20,333.0	8,268.4	(12,064.6)
France	21.1	114.6	93.5
Germany, West	4,797.3	582.5	(4,214.8)
Italy	169.1	22.3	(146.8)
Japan	22,323.0	521.9	(21,801.1)
Korea, South	1,070.2	67.6	(1,002.6)
Mexico	2,775.9	220.0	(2,555.9)
Sweden	1,128.8	30.6	(1,098.2)
United Kingdom	520.2	62.4	(457.8)
Others	38.0	3,062.8	3,024.8
Total	53,785.3	13,273.8	(40,511.5)

Data: Compiled from official statistics of U.S. Department of Commerce.
Source: MVMA, *Facts & Figures,* 1992, pp. 42–45.

of this, $40.7 billion (61 percent) was a direct result of auto imports, and the importation of Japanese vehicles accounted for over half, or $21.9 billion, of this amount. Several reasons have been advanced to explain this trade imbalance, particularly such macro-economic factors as the U.S. budget deficit, the dollar/yen exchange rate, and the low personal savings rate and low growth in productivity in the U.S.[6] While the automotive industry cannot take all the blame for the trade deficit, it nevertheless cannot avoid being a central trade issue since it accounts for such a large percentage of the deficit.[7]

The automotive industry is not the only sector that would benefit from increased domestic auto production—all related sectors would benefit. In 1990, for example, the automotive industry accounted for 13.1 percent of total steel materials, 15.8 percent of aluminum, and 50.0 percent of synthetic rubber consumed in the U.S.[8] It is also a major user of such "high-tech" items as computers, semiconductors, machine tools, and industrial robots. The automotive industry thus has a significant impact on many other sectors of the economy.

Can American manufacturers solve the current trade problem simply by coming up with measures to counter competitive pressures from foreign vehicle manufacturers? The answer, unfortunately, is "no." First of all, the demand for automobiles must be stimulated.

The number of new vehicles sold in the U.S. has remained flat for some time, even though the demand for cars in general has risen. As Figure 2.1 shows, the total number of vehicles in use (newly purchased and second-hand cars) has been steadily rising.

Figure. 2.1 Passenger cars: Sales (in million units) vs. cars in use (in 10 million units)

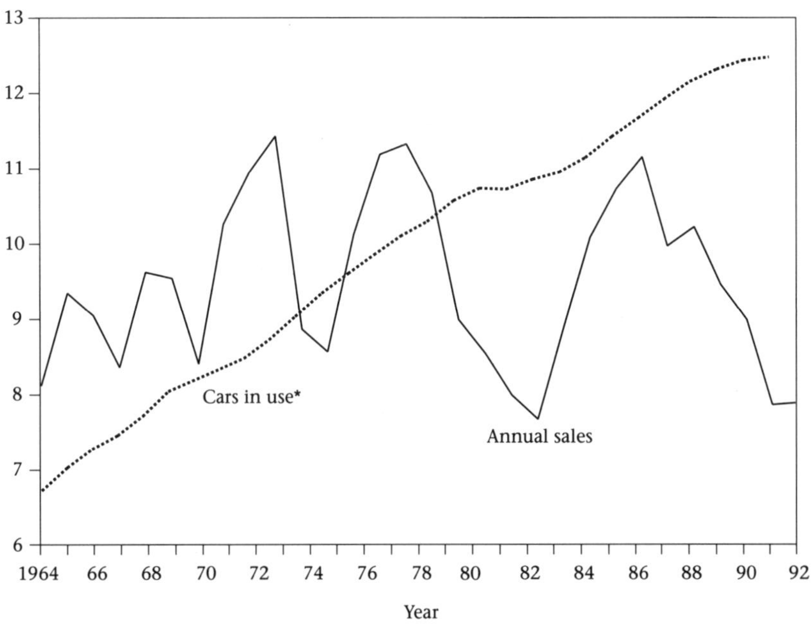

*Cars in use on July 1 of each year.
In 1979, 1.3 million window vans were transferred to the truck category.

Data: R. L. Polk & Co.; MVMA, *Facts & Figures*, various issues.

Source: *Ward's Automotive Reports*, March 30, 1992.

The demand for automobiles, on the whole, has been rising, yet new auto sales have been declining. Why? Because consumers are hanging onto their cars longer than they used to. As Figure 2.2 indicates, the average age of passenger cars and trucks is rising. Although new car sales rose in the latter half of the 1980s, older vehicles continued to be used, thus pushing up the average age of the vehicles in use.

The appeal of the new car seems to have dwindled in

Figure. 2.2 Average age of vehicles in use

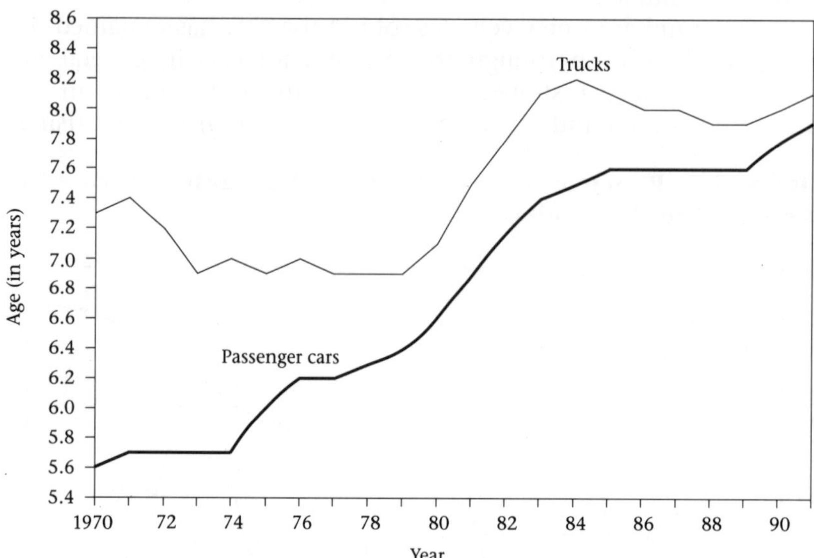

Data: R. L. Polk & Co.

Sources: MVMA, *Facts & Figures*, various issues.

comparison to that of the used car. Why? What can be done to reverse this trend, to stimulate demand for new cars? This is a key question this book will pursue. The author hopes to provide some practical ideas for restructuring the U.S. automotive industry.

Notes

1. U.S. Department of Commerce, "Key source data and assumptions for the advanced estimates of GNP: Easier access and redesigned format," *Survey of Current Business*, July 1988, pp. 128–30.
2. The U.S. lost its lead in auto production to Japan in 1980. Although American production increased significantly in the boom years of 1984 and 1985, U.S. production never again matched that of Japan.
3. The Free Trade Agreement (FTA) between the U.S. and Canada allows both countries to export/import vehicles duty-free. The Big Three trade vehicles between plants located in the two countries, and these are normally counted as domestically produced vehicles (Robert M. Stern, Philip H. Trezise, and John Whalley, eds., *Perspective on a U.S.-Canadian Free Trade Agreement*, Washington, DC: Brookings Institute, 1987).

4. Defining the term "imported vehicle" is a complex issue. Some vehicles made in Japan are sold under Japanese brand names; others are captive imports sold under American brand names. For example, of the 1.5 million "imported cars" assembled in Japan in 1991, over 10 percent were marketed as Big Three models ("U.S., Japan makers deeply intertwined," *Ward's Automotive Reports*, February 24, 1992, p. 1). There is no clear definition of "imported cars," but most statistics use figures that include captive imports.
5. This is in spite of flat truck sales that resulted from the substantial increase (from 4 to 25 percent) in import duty on chassis-with-cab for pickup trucks. In 1991, the Big Three charged that Toyota and Mazda were dumping their minivans onto the U.S. market. On May 19, 1992, the U.S. Department of State determined that these were indeed instances of dumping. The discounts offered were: Toyota, 6.75 percent; Mazda, 12.70 percent; and others, 9.88 percent. The Japanese manufacturers argued, to the International Trade Committee (ITC), that the prices of parts used in judging whether the vehicles had been dumped onto the U.S. market were determined without rational basis. On June 24, 1992, the ITC decided by a vote of four to two that domestic auto manufacturers had not been harmed by Japanese minivan dumping.
6. For a further discussion of these factors, see Lester Thurow, *Head To Head*, New York: Morrow, 1992; or Paul Krugman, *The Age of Diminished Expectations*, Cambridge, MA: MIT Press, 1990.
7. For a detailed account of the research that has been conducted in the area of trade imbalance in the auto industry, see Sean P. McAlinder et al., *The U.S.-Japan Automobile Bilateral 1994 Trade Deficit*, Office for the Study of Automotive Transportation, Transportation Research Institute, University of Michigan, May 1991. The Japan Automobile Manufacturers' Association has published a detailed argument refuting the findings of this report, questioning the methods employed in estimating the size of the market among other issues. The point to be noted here is that the trade imbalance has attracted so much attention that reports of this kind are now available.
8. MVMA, *Facts & Figures*, 1991.

   Motor Vehicles Manufacturers Association of the United States, Inc. (MVMA) was reorganized as American Automobile Manufacturers Association (AAMA) after December 1992. The members of AAMA are the three U.S. manufacturers: GM, Ford and Chrysler. The former members of MVMA such as Honda of America Mfg., Inc., Navistar International Transportation Corp., PACCAR Inc., and Volvo North America Corporation are excluded.

3

An Overview of Past Analyses

Obviously, this author is not the first to attempt an in-depth analysis of the U.S. automotive industry. Numerous excellent books and papers have focused on the various aspects of the industry, and an understanding of the important issues that they have examined is fundamental to gaining a broad-based perspective of the industry and its problems.

Before delving into the reasons behind the deterioration of the U.S. automotive industry in Part II, let us take a brief look at the issues identified by other authors. We will herein attempt to place the situation in its proper perspective by summarizing the issues that other authors have already pointed out.

This chapter spotlights five major problem areas that have been analyzed by other authors, that is, problems pertaining to management, technology, labor unions, automobile styling, and regulatory issues.

Management Problems

Many of the problems that confront the automotive industry are directly linked to the corporate culture and management concerns. Maryann Keller, for example, highlights in *Rude Awakening*[1] the obstructive elements within General Motors' corporate culture: bureaucratic constraints prevent change, and board members dare not disagree with the chairman. Individual effort is too often directed not at improving profitability or operating results, but rather toward climbing to the next rung on the corporate ladder.

> Once an executive reaches a level of prestige at General Motors, he hangs on for dear life. A visible change is likely to be seen in his demeanor. Gradually, he stops seeing the company's flaws and begins to develop a defensive posture toward critics and skeptics. Conformity is required, but an unbridled ambition also dictates his behavior.[2]

This type of secretive atmosphere makes it difficult to find (or even to look for) fundamental solutions, and it impedes collective bargaining with the UAW. Keller's analysis makes it clear that continued procrastination by GM in dealing with real issues will weaken the company not only vis-a-vis Japanese automakers, but also in relation to its U.S. rivals, such as Ford.

In the 1980s, the strong personality of Roger Smith determined the course of GM. In *Call Me Roger,*[3] Albert Lee provides an interesting account of the background against which GM pursued its strategies in the 1980s.

Management-related problems are not restricted to GM; lax awareness and deterioration of morale are rife within Ford and Chrysler as well. For all of the Big Three, these problems can be directly blamed on their enormous size.

David Halberstam (*The Reckoning*)[4] and Michael Moritz and Barret Seaman (*Going for Broke*)[5] take a documentary approach to their subjects. Their books depict, on an almost real-time basis, how business deteriorated for the Big Three. Although the authors are immensely persuasive in this respect, they nevertheless fail to pinpoint the fundamental problems that the leadership of these companies ought to be tackling.

That there is a crisis is evident to all. And it is correct to point out that countermeasures need to be taken immediately. But what countermeasures? Perhaps because the management of the Big Three were not totally clear on this point either, they had no alternative but to lag behind foreign competitors.

Technology Problems

Two books, entitled *The Machine that Changed the World*[6] and *The*

Future of the Automobile,[7] are part of a series of reports on the automotive industry published by the Massachusetts Institute of Technology (MIT). The former publication is well-known for having proposed the concept of "the lean production system."

The "Japanese method of production," signified by such procedures as Toyota's *Kanban* or "Just-In-Time" method, eliminated all possible waste on the production line. The methodology helped to enhance Toyota's competitiveness in the international marketplace. In the U.S., the Ford Production System was a methodology that enabled manufacturers to mass-produce a limited variety of vehicles.

> The lean producer, by contrast, combines the advantages of craft and mass production, while avoiding the high cost of the former and the rigidity of the latter. Toward this end, lean producers employ teams of multiskilled workers at all levels of the organization and use highly flexible, increasingly automated machines to produce volumes of products in enormous variety.[8]

A survey of more than ninety auto plants in seventeen countries revealed that Japanese automakers have achieved high standards of productivity and quality. Among the U.S. manufacturers, only the productivity at Ford is comparable to that of the Japanese manufacturers; the others are 50 percent lower, on average, than their Japanese counterparts. The Europeans scored lowest in the area of productivity. Japanese transplants in the U.S. are now able to produce cars of similar quality to Japanese-made ones, but the productivity in these transplants is 25 percent lower than in Japan.

The conclusion is that the American automakers face the need to transform themselves into "lean producers" in order to regain their competitive capabilities. This will require financial support, and the manufacturers will have to change their business perspective. These are points well taken; in fact, GM seems to have positioned itself toward achieving a more flexible system of production in the future.[9]

William J. Abernathy et al. point out the problems pertaining to production methodology in *Industrial Renaissance.*[10] This "classic" compares production in Japan and the U.S., focusing on the issue of cost.

The troubles that the U.S. automakers faced in 1991 may have come about as a result of not having had enough time to rectify their technological shortfalls. Although they may have been aware of their problems in this area, it is possible that the cyclical recession came around before they were able to take the necessary steps. Some Japanese auto industry experts have pointed out that the U.S. auto manufacturers will be able to achieve the same level of efficiency as the Japanese by the end of this decade as a result of measures that are currently being taken.

As discussed later in this book, it was only up to the early 1980s that American auto manufacturers lagged behind the Japanese in terms of price competition. In the latter half of the decade, the Japanese raised the prices of their vehicles at a faster rate than their American counterparts. It remains doubtful, though, whether the U.S. automakers are capable of enhancing their profitability solely through reduced production costs.

Labor Union Problems

Paul Krugman asserts in "The U.S. response to foreign industrial targeting"[11] that high wages were the main cause of the trade imbalance. Table 3.1 compares the hourly wages of workers in the automotive industry with those in the manufacturing sector. In 1991, the hourly wages earned by production workers making motor vehicles and car bodies (Standard Industrial Classification SIC 3711) averaged $21.30, or 1.9 times the $11.50 earned by workers in the manufacturing sector. Production workers manufacturing motor vehicles and vehicle parts and accessories (SIC 371) earned $15.30. Workers hired by the Big Three earn wages that are much higher than at other automakers.[12]

Krugman has examined what happens to an industry that pays such high wage rates when price competition arises from foreign competitors. He asserts that the industry will lose domestic market share because it lacks competitive capability; the industry is eventually forced to reduce the number of workers. On the other hand, consumers will benefit from lower prices. The issue boils down to which is of more importance: the cost of the job cuts or the consumers' surplus. Krugman concludes that if wages were to be maintained at high levels due to pressure from the labor union, a part of the consumers' surplus will flow to foreign manufacturers; thus, the reduction in income caused by job cuts will be greater than the consumers' surplus.

Consumers are thus negatively affected by the higher wage rates paid in the auto industry. When cheap imports flow into the country under these circumstances, the loss of income brought about by job cuts expands the negative effect to the overall

Table 3.1 Wage rate comparison: Hourly earnings (in dollars) of production workers

Year	All manufacturing Payroll (1)	Motor vehicles and equipment (SIC 371) Payroll (2)	Motor vehicles and car bodies (SIC 3711) Payroll (1)	Motor vehicles and car bodies (SIC 3711) With other benefits* (1)	GM With other benefits (3)
1980	7.4	9.85	12.6	17.9	18.5
1981	8.1	11.02	13.9	19.0	19.8
1982	8.7	11.62	14.4	20.4	21.5
1983	9.0	12.14	14.8	20.0	21.8
1984	9.4	12.73	14.5	19.1	22.6
1985	9.9	13.39	16.7	21.9	23.4
1986	10.2	13.45	17.2	22.8	24.0
1987	10.3	13.57	17.3	23.3	25.9
1988	10.7	13.90	18.7	24.9	27.9
1989	10.9	14.28	19.4	26.4	29.5
1990	11.2	14.59	20.3	28.3	31.3
1991	11.5	15.32	21.3	30.8	34.6

*Payroll + (social security and other legally required payments + employer payments and other programs), estimated by K. Ohmura.

Sources: (1) U.S. Department of Commerce, Bureau of Census, *Annual Survey of Manufactures*, 1980, 1981, 1983, 1984, 1985, 1986, 1988, 1989, 1990, 1991; *Census of Manufactures*, 1982, 1987.
(2) MVMA, *Facts & Figures*, 1992, p. 62.
(3) General Motors, *Annual Report*, various years.

economy, a rather ironic ending. After pointing out that wage rates are higher in the auto and steel industries than in other manufacturing sectors, Krugman says:

> In sum, then, foreign competition in autos and steel has actually or, without protection, potentially displaced U.S. workers from high wage jobs, and in these cases at least some of the wage loss represents a net cost to the U.S. economy rather than simply a redistribution from one set of U.S. residents to another. This downgrading of workers, however, is essentially limited to the auto and steel industries and is not typical of the economy as a whole: and there is no good case for arguing that foreign targeting has been important in the loss of competitiveness of autos or has been a major factor in steel.[13]

Further studies are required on the role of the labor unions in the determination of wage rates. Already a number of managers assert that high wages achieved by the United Automobile Workers (UAW) pose a problem for enhancing corporate competitive capability. The issue of high wages will play an important role in Part IV of this book, where it will be examined from a different angle, that is, from the viewpoint of high wages being synonymous with high income.

Style Problems

Probably the most general criticism that is made of American auto manufacturers is that they do not produce vehicles that are affordable, that their cars break down often, and that the style is not appealing. The issue has become a staple joke in magazines, comic books, and television shows.

Can it really be possible that the world's leading manufacturers have actually failed to develop attractive vehicles for twenty years? There must be an explanation for this. Toward this end Clark et al. analyze the research and development activities of the major auto manufacturers around the world in the report entitled "Product development in the world auto industry."[14] This report offers an objective benchmark in the area of research and development, which is an aspect that is very often difficult to quantify.

In terms of the man-hours put into developmental work by engineers, a Japanese engineer spends 55 percent of the time spent by an American engineer, and just 42 percent that of a European engineer. The difference is due primarily to the role played by parts manufacturers. U.S. firms have recently begun to actively introduce the concept of "design in," which involves parts manufacturers from the initial developmental stage.

Fred Mannering and Clifford Winston[15] provide interesting insights into American consumers' loyalty to American brands. It takes more time to regain brand loyalty than to regain quality and price competitiveness. The authors point out that this was the major reason behind the deterioration of market share held by the American manufacturers in the latter half of the 1980s. Conclusions arrived by means of econometric methodology indicate that consumers' loyalty to American brands have deteriorated over the past twenty years.

Passenger cars, especially, rely on the consumers' brand

loyalty. Manufacturers gain a non-price competitive edge once they have established brand loyalty among their customers; Mercedes Benz and Jaguar are prime examples of this. However, it is doubtful that such loyalty can be achieved for all product lines. Consumers are very price conscious when it comes to commuter cars and entry-level vehicles for first-time drivers. Further studies need to be conducted to determine whether the issue is that of American consumers forsaking American vehicles because of the perception that they offer lower performance and quality at higher prices, or whether American vehicles truly do lag behind their foreign rivals in these areas.

Regulatory Issues

One theory asserts that governmental regulations on corporate average fuel efficiency (CAFE), emission control, and safety are placing an extra burden on the automotive manufacturers. In *Regulating the Automobile,*[16] Robert W. Crandall et al. make a claim that is representative of this view.

The authors estimate that consumers were forced to pay an extra $1,300 to $2,200 per vehicle to purchase automobiles that complied with federal safety and emission control regulations of 1981. This proved to be a burden on consumer purchasing power and led to a subsequent decline in new auto sales.[17]

The authors also point out that the credibility of American vehicles among consumers declined with the heightening of regulatory controls. They argue that regulations should be implemented from a long-term perspective and be of the kind that allow manufacturers enough room to maneuver and respond to changing circumstances. Further studies need to be done as to why foreign vehicles were able to comply with tougher regulatory controls at lower cost.

Similarly Lawrence J. White points out in *The Automobile Industry since 1945*[18] that antitrust regulations have restrained GM. His book focuses on GM's monopolistic presence in the auto market up until the 1960s, and the need to curtail such a market force. Though dated, this book continues to be valuable after all these years because it offers the reader a comprehensive background to the auto industry.

Notes

1. Maryann Keller, *Rude Awakening,* New York: Harper Perennial, 1989.
2. Ibid., p. 17.
3. Albert Lee, *Call Me Roger,* Chicago: Contemporary Books, 1988.
4. David Halberstam, *The Reckoning,* New York: Morrow, 1986.
5. Michael Moritz and Barret Seaman, *Going for Broke: The Chrysler Story,* New York: Doubleday, 1981.
6. James P. Womack, Daniel T. Jones, and Daniel Roos, *The Machine that Changed the World,* New York: Rawson Associates, 1990.
7. Alan Atshuler et al., *The Future of the Automobile,* Cambridge, MA: MIT Press, 1984.
8. Womack et al., p. 13. (Reprinted with permission of Rawson Associates, an imprint of Macmillan Publishing Company from *The Machine that Changed the World* by James P. Womack, Daniel T. Jones, and Daniel Roos. Copyright © 1990 James P. Womack, Daniel T. Jones, Daniel Roos and Donna Sammons Carpenter.)
9. "Platform madness," *Forbes,* January 20, 1992, pp. 40–41; "Follow that Ford," *Forbes,* April 27, 1992, p. 44; "General Motors: Open all night," *Business Week,* June 1, 1992, pp. 66–67.
10. William J. Abernathy, Kim B. Clark, and Alan M. Kantrow, *Industrial Renaissance,* New York: Basic Books, 1983.
11. Paul Krugman, "The U.S. response to foreign industrial targeting," *Brookings Papers on Economic Activity,* Vol. 1, 1984, pp. 77–131.
12. For the sake of international comparison, the following breakdown was reported for 1988:

Table 3.2 Motor vehicle and equipment manufacturing production workers' hourly compensation in selected countries, 1988

	Total compensation (US$)	Percentage of U.S. compensation
U.S.	21.95	100
Canada	16.56	75
Japan	16.62	76
S. Korea	3.38	15
France	14.03	64
Germany	22.95	105
Italy	13.54	62
Spain	10.48	48
Sweden	17.12	78
U.K.	12.51	57

Source: MVMA, *Facts & Figures,* 1990, p. 71.

International competitiveness is not determined solely by wage differentials; productivity and the extent of automation are also important factors. However, it is self-evident that the U.S. would be unable to compete with Japan if Japanese wages are 75 percent of U.S. wages because Japanese productivity is higher. (The yen-to-dollar rate applied to Table 3.2 is 128 yen per dollar.)

13. Paul Krugman, p. 96.
14. Kim B. Clark, W. Bruce Chew, and Takahiro Fujimoto, "Product development in the world auto industry," *Brookings Papers on Economic Activity*, Vol. 3, 1987, pp. 729–81.
15. Fred Mannering and Clifford Winston, "Brand loyalty and the decline of American automobile firms," *Brookings Papers: Microeconomics*, 1991, pp. 67–114.
16. Robert W. Crandall, Howard K. Gruensprecht, Theodore E. Keeler, and Lester B. Lave, *Regulating the Automobile*, Washington, DC: Brookings Institute, 1986.
17. The average price of a new vehicle is $10,581 (in 1967 dollar terms), according to the MVMA. Of this, $2,643 is the cost that is added on to comply with safety and emission control standards (MVMA, *Facts & Figures*, 1991, p. 56).
18. Lawrence J. White, *The Automobile Industry since 1945*, Cambridge, MA: Harvard University Press, 1971.

PART II

From Heaven to Hell

4

Stable Demand, Deteriorating Output

Let us begin our search for solutions to the issues facing the automotive industry by briefly examining the historical background that led to the current decline of the U.S. automakers. When did the automotive industry begin to weaken? How did it deteriorate? What aspects of it deteriorated specifically? Are the characteristics of the recession of 1990/91 similar to those of past recessions, or has the economic environment deteriorated even further? These are the main issues that we shall address.

Figure 4.1 shows vehicle output as a percentage of the U.S. GNP since 1967.[1] The output shown in this chart is not restricted to that of the Big Three automakers. Rather, it is the total amount of new and used passenger cars and trucks produced and traded (including the value of the vehicles produced not only by the Big Three but also by the Japanese and German transplants, while excluding vehicles imported from Canada and Mexico by the Big Three and the Japanese but sold under the brand name of the Big Three and Japanese).

The first thing made evident by this chart is the fluctuation in output. There seems to have been a four- to five-year cycle from the late 1960s through the 1970s. In contrast, the industry weakened over the three-year period from 1980 to 1982; after peaking in 1983, output leveled off and the cycle has become longer.

Figure 4.1 Vehicle output as percentage of GNP

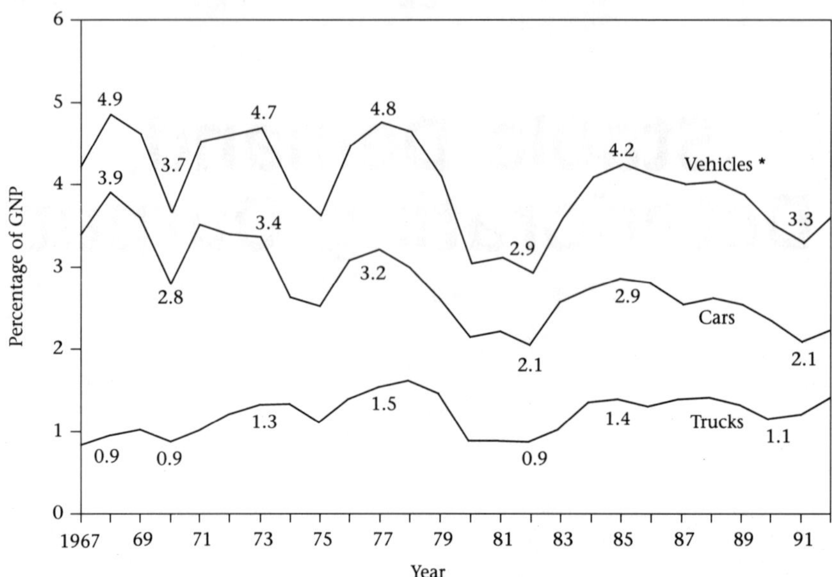

*Vehicles = Cars + Trucks

Sources: MVMA, *Facts & Figures*, 1992; U.S. Department of Commerce, "National income and product accounts," *Survey of Current Business*, February 1993.

If cyclic elements are ignored, the chart indicates a gradually declining trend in vehicle output. The total output of passenger cars and trucks apparently fluctuates within a fixed range until the late 1970s, then hits its nadir during the recession of 1980/82 (the severest recession ever experienced by the U.S.). The recession of 1990/91 was also deep, comparable with the previous one. It seems probable that something which took place in the latter half of the 1970s led to the change seen in the subsequent decades.

If we examine passenger car output and truck output separately (see Figure 4.1), we see a clear downward trend for passenger car output. The output of trucks, however, rose during the 1970s before showing a drop through the recessionary period in the 1980s, similar to the situation for passenger cars. Truck output peaked in 1985 at a level near that of 1978 (1.6 percent of GNP).

Thus, as was suggested in Part I, the problems are more evident for passenger cars. We will hereafter focus on passenger cars, therefore, in our attempt to ascertain the reasons behind the long-term decline of the American automotive industry.

The question we will seek to answer is whether the decline in the output of vehicles as a percentage of GNP was due to reduced demand, or whether it was a direct consequence of a change in the structure of automotive production. In an advanced economy such as the U.S., consumption tends to shift away from material goods; in other words, the economy becomes more "service" oriented. To attain a convincing answer, therefore, we must examine whether the declining demand for new automobiles has been in line with such a trend. If consumer preference is for travel by public transport, such as by plane rather than by car, then auto-related expenditures (as a percentage of total personal expenditures) would naturally decline.

Another important phenomenon is the shift away from manufacturing. If a nation becomes more dependent on imported goods (for example, as a result of decreased competitive capability in the international marketplace), this leads to a retrenching of its manufacturing sector. It is highly likely that this has taken place in the U.S.—since its automotive industry pays wage rates that are much higher than in other industries—as evidenced by the fact that imported vehicles account for a higher proportion of the vehicles sold. A detailed examination will be required, however, to determine whether this is the sole reason behind the decline in domestic vehicle output.

Table 4.1, which shows the demand for cars (the dollar amount of cars sold), dollar amount of cars manufactured, and net imports (imports minus exports), offers a good bird's-eye view of whether the phenomena are actually taking place. The difference between cars manufactured and domestic demand implies net imports in the long-term. If the problem lies with demand, we should see falling demand; if with production, we would expect to see vehicle output falling faster than decline in demand, and imports should increase.

Table 4.1 clearly shows that the problem lies with the structure of production. There is no significant difference between domestic demand and output until the 1970s. The ratio of imports (that is, net imports divided by domestic demand) fluctuated between 0.4 percent and 1.5 percent in the first half of the 1960s, but rose to between 2 percent and 9 percent in the latter half of the decade. In the first half of the 1970s, the ratio reached 8.8 percent through 11.6 percent (owing in part to a shortage of supply caused by striking workers at GM). In the latter half of the decade, the ratio hovered between 9.6 percent and 13.0 percent. The gradual penetration of imported cars over the years is obvious.

Table 4.1 Demand, output and net import of automobile sector (billions of dollars)

Year	Current domestic demand	Output	Net import	% import/ demand
1960	20.7	21.3	0.3	1.5
1961	18.6	17.8	0.1	0.4
1962	22.5	22.5	0.2	0.8
1963	24.9	25.1	0.2	0.6
1964	26.3	25.9	0.2	0.9
1965	30.7	31.2	0.3	0.9
1966	30.1	30.2	0.7	2.2
1967	29.4	27.8	0.9	3.1
1968	35.6	35.0	1.8	5.1
1969	36.9	34.8	2.4	6.4
1970	32.5	28.5	2.9	8.8
1971	41.5	38.9	3.9	9.5
1972	46.0	41.4	4.4	9.6
1973	49.8	45.9	4.7	9.4
1974	42.6	38.8	5.0	11.6
1975	46.1	40.3	4.3	9.3
1976	59.5	55.1	5.7	9.6
1977	70.0	64.2	7.0	10.0
1978	77.5	67.9	10.0	12.9
1979	77.8	66.2	10.1	13.0
1980	72.9	59.2	12.8	17.6
1981	81.7	68.3	13.8	16.9
1982	85.1	65.3	17.4	20.4
1983	105.9	88.4	19.3	18.2
1984	126.8	104.2	25.8	20.3
1985	141.4	115.8	30.1	21.3
1986	158.0	120.4	39.0	24.7
1987	152.7	118.9	40.6	26.6
1988	166.1	129.1	38.1	22.9
1989	165.9	133.9	33.7	20.3
1990	170.4	130.3	35.4	20.7
1991	153.6	119.7	33.9	22.1
1992	166.3	133.2	32.8	19.7
Compound growth rate 1992/1960	6.73%	5.89%	15.80%	

Sources: U.S. Department of Commerce, Bureau of Economic Analysis; U.S. Department of Commerce, "National income and product account," *Survey of Current Business*, February 1994, p. 27.

In the first half of the 1980s, the ratio began to rise sharply, to between 16.9 percent and 21.3 percent; it rose even further in the latter half of the decade, to between 20.3 percent and 26.6 percent. The ratio peaked in 1987, when the value of the dollar plunged against foreign currencies, which enhanced the relative competitive capability of U.S. exports in the world marketplace.

Were imports to blame for the decline in car output around 1980? It should be noted here that foreign autos were freely imported only until 1980/81, when Chrysler faced potential bankruptcy. Japanese manufacturers voluntarily restricted their exports to the U.S. after that time.

The Big Three were still in good shape in 1986/87, when the proportion of imported cars hit its peak (see Table 1.1). Imported cars were not then a significant issue. The number of imported cars actually declined in the latter half of the 1980s, when the Japanese switched from exporting assembled vehicles to raising the production capacity of their transplants (see Table 2.4). Though the penetration of imported cars had become a significant issue, it does not seem to have had much impact on the recession of 1990/91.

Figure 4.2 indicates domestic passenger car demand and output as a percentage of the GNP. The line indicating output corresponds to the one used in the previous chart, but an interesting point may be observed here in contrast to domestic demand.

In comparison to the 1970s, output is clearly down in the 1980s. Domestic demand, however, reached 3.7 percent of GNP in 1986, which was nearly as much as the peak of 3.8 percent witnessed in 1971/72 (and higher than the 3.5 percent posted in 1977). Demand bottomed out at 2.7 percent in 1980/82 and again in 1991, lower than the previous 2.9 percent low witnessed in 1974/75.

The numbers for demand are thus similar for both decades. The decline witnessed in 1990/91 was caused by a cyclical deterioration in the economy, not by a structural decline in domestic demand itself.

There seems to be no special factor differentiating the recession of 1990/91 from that of 1980/82. A plausible explanation is that the recent recession was a cyclic phenomenon of the economy, which is also true of the previous recession. In other words, we do not see a sudden rise in imports, and domestic demand did not suddenly evaporate. In the following chapters, we will examine the various factors that affect domestic demand and output.

Figure 4.2 Automotive demand and output as percentage of GNP

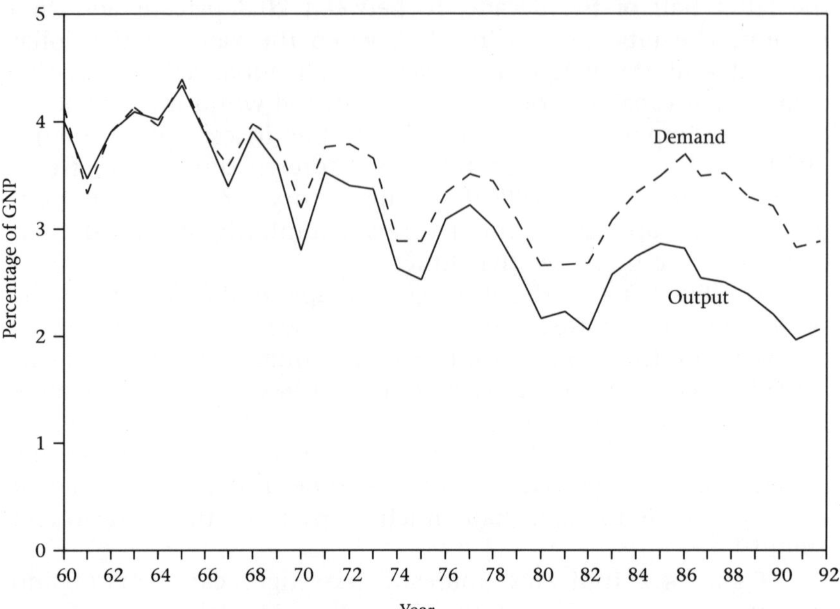

Sources: MVMA, *Facts & Figures*, various issues; U.S. Department of Commerce, "National income and product accounts," *Survey of Current Business*, February 1994.

Note

1. Unless otherwise noted, the statistics in this chapter are based on nominal prices. The prices that form the basis for value-based statistics are "actual transaction prices," which fluctuate independent of the Consumer Price Index (CPI, used to deflate the nominal series into the real series).

   The GDP and automotive statistics in this chapter are used in reference to the National Income and Product Account. Statistics for the automotive industry are based on units of vehicles; in most cases, value-based statistics are not required other than when relating them to the overall economy. The following points should be noted when dealing with value-based statistics:

   - The most fundamental statistic is value of final sales, which is obtained by multiplying the number of vehicles sold by average (actual transaction) price. Prices are estimated on the basis of invoice data and take into consideration the over-valuation of trade-ins for used cars and discounts offered by manufacturers and dealers (that is, they differ from the suggested retail price set by the automakers).

- Sales amounts can be broken down into personal consumption expenditure, producers' durable equipment, and government purchases. Adjustment is made for net imports and change in business inventories to obtain the output amount (that is, output value = domestic sales (personal consumption + producers' durable equipment + government purchase) − net imports ± changes in business inventories).
- Special attention must be given to the statistics that include used cars. The value of used car sales is derived by multiplying the number of units sold by the margin. Since in the exchange of a used car there is no manufacturer, in terms of the national economy, the only output of the used car business is the commission received by the dealers.

For further details on the relevant statistics, refer to the following articles in *Survey of Current Business*:

U.S. Department of Commerce, "Automobile purchase by business and consumers as reflected in the national accounts," March 1962, pp. 13–24.

U.S. Department of Commerce, "Vehicles: Recent developments and treatment in the GNP accounts," November 1975, pp. 4–7.

Carol S. Carson, "GNP: An overview of source data and estimating methods," July 1987, pp. 103–27.

Robert P. Parker, "A preview of the comprehensive revision of the national income and product accounts: New and redesigned tables," October 1991, pp. 20–27.

5

The Waning Appeal of the New Car

Automobiles can be regarded as "glamor" products. They signify an era and provide the background for a historical moment. Automobiles tantalize our aspirations, as exemplified by such cars as the Cadillac, Lincoln, Mustang, Cougar, Accord, Taurus, and Lexus. Certain cars can provide a fleeting sense of delight in a movie scene, in town, or on the expressway, causing us to fantasize, "Gee, I'd like to drive that car!" In short, selling cars is a drama that extends beyond a mere business.[1]

Consumer taste is strongly reflected in demand; a significant number of people want to purchase or trade up to a car that both fulfills their needs and satisfies their preference for style. A particular model of car can easily become a hit product if it meets this potential demand by offering a new design or technology. This in turn can have a significant effect on the market shares held by the individual automakers.

Product appeal is undeniably important in the automotive industry; unfortunately, it is difficult to quantify and virtually impossible to accurately measure. The need for arduous product development would be eased if a formula could be drawn up to quantify the prospective appeal of the proposed model. The very fact that this has not yet been achieved, and indeed may never be realized, offers potential to the industry to add value in this area.

The important question in this regard is: Does personal

expenditure on automobiles increase if a very popular car appears on the market? Will consumers loosen their purse strings if an attractive product is offered? The answer to these questions, unfortunately, is "no."

Consumers tend to confine their auto-related expenditures to a narrow range. Table 5.1 shows the proportion of spending

Table 5.1 Personal consumption expenditures for transportation as percentage of total personal consumption

Year	New autos	Net purchase of used autos	Other vehicles*	Total autos	Repair, greasing, washing, parking, storage, and rental	Gas and oil	Other expenses	Total
1966	4.5	0.9	0.3	5.7	1.6	3.4	2.4	13.1
1968	4.6	0.9	0.4	6.0	1.7	3.4	2.3	13.4
1970	3.6	0.9	0.4	4.9	1.8	3.6	2.3	12.6
1971	4.3	1.0	0.6	5.8	1.9	3.5	2.4	13.6
1972	4.2	0.9	0.7	5.8	1.8	3.2	2.3	13.1
1973	4.2	1.0	0.7	5.9	1.9	3.5	2.3	13.7
1974	3.1	1.0	0.5	4.6	2.0	4.1	2.2	12.9
1975	3.1	1.0	0.6	4.7	2.1	4.0	2.0	12.8
1976	3.5	1.1	1.0	5.6	2.0	4.1	2.7	14.3
1977	3.7	1.1	1.1	5.9	2.0	4.0	2.9	14.8
1978	3.6	1.1	1.2	5.9	2.1	3.9	2.8	14.7
1979	3.3	1.1	0.9	5.2	2.1	4.5	2.8	14.6
1980	2.7	0.9	0.7	4.2	1.9	4.8	2.8	13.8
1981	2.7	1.0	0.6	4.2	1.9	5.3	2.8	14.1
1982	2.6	1.0	0.8	4.3	1.7	4.3	2.7	13.1
1983	3.0	1.2	0.9	5.1	1.8	4.2	2.7	13.7
1984	3.2	1.2	1.1	5.5	1.8	3.7	2.6	13.6
1985	3.2	1.0	1.1	5.3	1.7	3.3	2.4	12.7
1986	3.6	1.3	1.2	6.1	1.9	2.6	2.5	13.1
1987	3.1	1.3	1.2	5.5	1.8	2.5	2.6	12.5
1988	3.1	1.3	1.2	5.6	2.0	2.4	2.6	12.6
1989	2.9	1.2	1.2	5.3	2.0	2.4	2.6	12.3
1990	2.6	1.2	1.1	4.9	2.1	2.6	2.6	12.1

*New and used trucks, recreational vehicles, and the like.

After 1991, the definition of "used cars" has changed. Although the consistent series of new data are only available from 1982, the trend that consumers' expenses for new cars have been decreasing while those for used cars and other vehicles have been increasing has been shown to be equal to the old series.

Data: U.S. Department of Commerce, Bureau of Economic Analysis.

Sources: MVMA, *Facts and Figures*, various issues.

allocated to automobiles in comparison to overall personal spending. We see that the expenditure for passenger cars, used cars, and other vehicles fluctuates between 4.2 percent and 6.1 percent of overall personal spending. If a new passenger car induces consumers to spend more to purchase that car, the result is reduced spending on used cars and other vehicles. The demand for automobiles is thus directly dependent on income growth.

Within the past twenty-five years, the expenditure for autos reached its peak in 1986. This was due, however, not to increased spending on new cars, but to a rise in purchases of used cars and other vehicles. The percentage of total personal consumption allotted to the purchase of used cars rose to 1.3 percent in 1986. Purchases of other vehicles (such as trucks or recreational vehicles) showed a rising trend and nearly tripled to 1.2 percent from the 0.4 percent posted around 1970.

Consumers allot a fixed proportion of their spending to automobiles, suggesting that the demand for automobiles should rise in line with growth in income (which implies stepped-up spending). The logic is simple enough. In fact, personal spending as a percentage of the U.S. GDP has risen over the past twenty years. The purchasing power of households (and their interest in obtaining cars) has been rising.

The proportion of personal spending as a percentage of the GDP rose conspicuously during the latter half of the 1980s, when the U.S. economy went through an expansive phase. Reaganomics stimulated personal spending by means of personal tax cuts and disinflation brought about by the fall in oil prices. The U.S. experienced the longest period of expansion since World War II, driven by strong personal spending.

The shift to demand for services rather than material goods, which normally accompanies a maturing economy, has been slow. Although the proportion of personal durable consumption as a percentage of personal spending declined in the latter half of the 1980s, its level during the recession of 1990/91 was similar to that of the recession in 1980/82. Thus, we find no special clues pointing to a structural decline in the demand for durable goods.

What did deteriorate during the recession of 1990/91 in comparison to that of 1980/82 was the demand for new cars. Although 1991 was the severest year in terms of profitability of the Big Three manufacturers, as explained in Chapter 1, spending allotted for the purchase of automobiles has remained relatively consistent (although there has been a rapid decline in the proportion of new car sales). This trend is readily apparent in

Figure 5.1. During the latter half of the 1960s, 80 percent of the passenger cars purchased were new cars. Used cars and other vehicles gradually began to take over, however, and the ratio of new cars had fallen to 53 percent in 1990. Obviously, new auto sales is a critical factor affecting car output—active trading in used cars does not help the automakers' business.

Figure 5.1 Personal consumption: New cars, used cars, and other vehicles

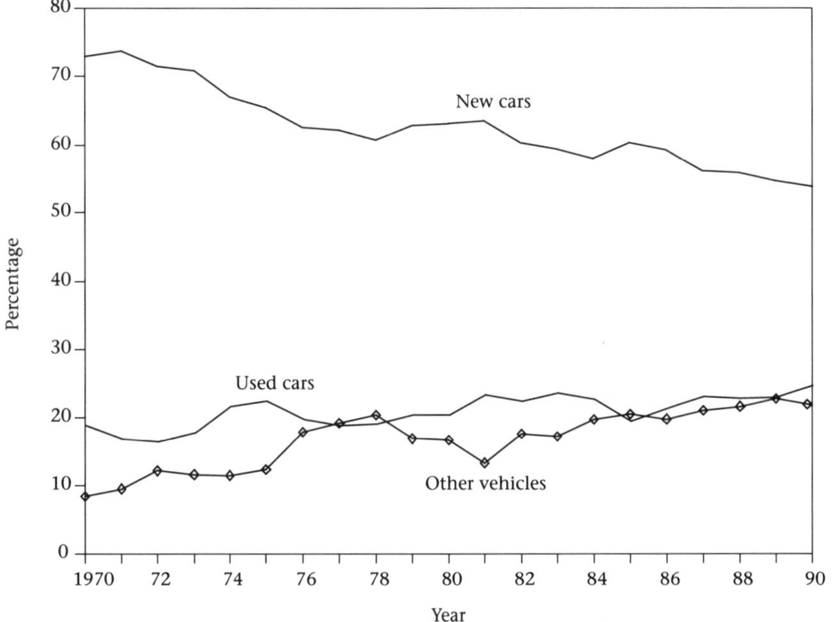

See the footnotes of Table 5.1.

Source: Department of Commerce, Bureau of Economic Analysis.

Let us pause here to summarize the issues identified thus far:

- The demand for automobiles has been maintained at a fixed level of total personal expenditure.
- Demand for new cars has fallen off.
- Demand for used cars has been rising.

These issues are in line with those alluded to in the previous chapter. Weaker demand for passenger cars is the main cause of dwindling earnings posted by the Big Three. While the number of new cars sold has decreased, however, the number of cars in use is on the rise; that is, the dependence on used cars has increased. The reason (on the demand-side) for deteriorating automotive sales during the recession of 1990/91 is the consumers' shift away from new cars.

The next topic that needs to be examined, then, is the reason behind this emerging preference for used cars over new cars. Who sells used cars to consumers? If a consumer sells a used vehicle to another consumer, the seller will own fewer cars. For instance, an owner of three cars may sell one of them, leaving him with only two cars. This does not reflect the current situation, however, because the number of cars in use is actually on the rise.

In fact, consumers are purchasing used fleet cars; that is, cars formerly purchased or leased for business use. As indicated in Table 5.2, consumer (rather than business) purchases accounted for 74 percent of total retail auto sales in 1970, but by 1990 this had declined to 60 percent. Although sales of cars to businesses has

Table 5.2 U.S. retail sales of new cars by sector (thousands)

Year	Consumer	% of total sales	Business	% of total sales	Total sales*	Number of cars in fleets of 10 or more	Number of cars in fleets of 4 or more
1970	6252	74.4	2056	24.5	8403	5041	9992
1975	5907	69.2	2508	29.4	8538	5956	10398
1980	6062	67.5	2791	31.1	8979	7143	10433
1981	5632	66.0	2787	32.7	8535	7130	10436
1982	5285	66.2	2593	32.5	7980	6861	10076
1983	6054	66.0	3006	32.7	9179	6932	10400
1984	6590	63.4	3669	35.3	10394	7217	10475
1985	7083	64.2	3822	34.6	11039	7600	10508
1986	7658	66.9	3666	32.0	11450	7756	10560
1987	6748	65.7	3395	33.0	10278	7934	10578
1988	6802	63.9	3699	34.8	10639	8201	10597
1989	6375	64.3	3402	34.3	9913	8318	10592
1990	5768	60.8	3567	37.6	9484	8312	10607
1991	4534	54.0	3758	44.8	8389	8076	10514

*Total sales is the sum of the sales to consumers, businesses, and the government.

Data: Bobit Publishing, *Automotive Fleet Book*, various issues.

Sources: MVMA, *Facts & Figures*, various issues.

increased, the fleet size owned by businesses has not. While large fleets (ten or more vehicles) are growing, the size of fleets of four or more vehicles has remained flat during the past twenty years.

The fact that the average fleet size has not expanded much indicates that businesses are getting rid of their used cars when making new purchases. If companies are buying more new cars for fleets, they must also be getting rid of a larger number of used cars.[2] A more precise description of the recent tendency would be that "households" are moving away from new cars.

It is relatively easy to explain why consumers are forsaking new cars. Table 5.3 explains the trend quite plainly: Consumer expenditure per new car has risen too much. Using 1960 prices as the index (that is, 100), gasoline and oil costs (per mile) rose to 255 in 1990 and maintenance costs to 278. The index price of new

Table 5.3 Index of passenger car operating costs

Year	Cost indices of selected items			Consumer expenditure per new car index (3)
	Gas and oil (1)	Maintenance (1)	Depreciation (2)	
1960	100.0	100.0	100.0	100.0
1965	98.5	86.1	96.9	105.6
1971	113.0	92.4	114.2	131.2
1975	184.0	122.8	119.7	173.5
1980	223.7	141.8	160.7	265.5
1981	239.3	149.4	199.2	312.3
1982	253.4	131.6	207.9	346.7
1983	236.3	131.6	186.8	372.9
1984	235.1	155.7	194.0	401.3
1985	171.0	173.4	204.3	421.4
1986	183.2	202.5	233.1	451.9
1987	198.5	202.5	276.2	478.7
1988	198.5	240.5	324.1	507.1
1989	206.1	265.8	364.9	536.0
1990	255.7	278.5	393.7	561.2
1991	229.0	278.5	430.3	580.4

Notes: (1) Based on costs in cents/mile.
(2) Vehicles for 1975 and later years are intermediates. Prior to 1974, full-sized vehicles were specified.
(3) Average retail price of all passenger cars including domestic and imported.

Sources: MVMA, *Facts & Figures*, various issues.

passenger cars has risen to 561. Depreciation cost (the cost of carrying cars) has also risen sharply to an index of 393. (The pricing policy of new cars is a central topic of this book and will be discussed in more detail in Part III.)

We have already seen that consumers allot a certain percentage of their spending toward cars. Thus, the sales of new cars must necessarily decline if prices rise. And if a family needs to own a certain number of cars, it has no choice but to switch from expensive new cars to the cheaper used ones.

The most significant characteristic of the recession of 1990/91 is that consumers gave up on the idea of buying new cars. This was a direct result of higher prices. If prices go up, demand falls—a simple rule of economics. In subsequent parts, we will examine the pricing strategy taken by auto manufacturers to determine why they resorted to such a strategy.[3]

Let us here step aside for a moment to examine the shift from large cars to smaller ones. Some researchers point to this as a major cause of the Big Three's decline. They suggest that the sudden rise in the popularity of fuel-efficient Japanese cars would not have taken place were it not for the two oil crises.

As indicated in Table 5.1, consumers allocate a certain proportion of their expenditure (that is, 12.1 percent to 14.8 percent) to auto-related spending. When gasoline prices shoot up, which is exactly what happened during the two oil crises, consumers face the following choices:

1. Reduce gas consumption by driving less.
2. Secure sufficient funds to pay for higher gas prices by reducing other types of spending.
3. Switch to a more fuel-efficient car.

Compare the data for 1972 and 1974 in Table 5.1. The percentage of spending on gas and oil rose from 3.2 percent in 1972 to 4.1 percent in 1974, which helps to explain the fall in expenditures for car purchases from 5.8 percent to 4.6 percent. Overall, consumers seemed to have opted for choice (3) above. As a result, when demand for new cars recovered between 1976 and 1978, smaller, fuel-efficient imports accounted for a larger percentage of the vehicles purchased (see Table 5.4).

During the second oil crisis, gas and oil expenditures rose once again, from 3.9 percent in 1978 to 5.3 percent in 1981. In response, the amount set aside by consumers for new car purchases declined from 5.9 percent to 4.2 percent during the same period. However, oil prices fell sharply after that.[4] As a result, the demand for

Table 5.4 Retail car sales by class (% of total sales)

Year	Sub-compact	Compact	Inter-mediate	Standard + Luxury	Import
1970	1.8	19.8	23.5	39.7	15.3
1971	7.4	15.8	20.8	40.8	15.2
1972	8.2	15.3	22.6	39.3	14.8
1973	10.7	16.8	23.4	33.7	15.4
1974	10.6	22.7	24.4	26.4	15.9
1975	12.3	22.9	23.9	22.6	18.3
1976	9.9	24.1	27.8	23.5	14.8
1977	8.5	21.2	26.4	25.4	18.6
1978	8.8	22.9	26.6	24.0	17.7
1979	12.9	20.9	23.8	20.6	21.9
1980	15.4	21.0	21.4	15.6	26.7
1981	17.1	20.2	20.4	15.1	27.3
1982	16.7	17.5	19.0	19.0	27.8
1983	12.7	16.4	26.5	18.4	26.0
1984	14.0	18.1	25.5	18.9	23.5
1985	14.0	17.1	26.4	16.8	25.7
1986	14.3	14.9	26.3	16.3	28.3
1987*	9.6	19.0	26.3	14.3	30.8
1988	9.3	19.7	28.6	13.5	28.9
1989	11.2	18.1	29.4	13.0	28.3
1990	11.8	20.7	27.6	13.3	26.6
1991	15.9	20.4	27.7	7.5	28.0

*Not comparable to previous data beginning in 1987.

Totals may not equal 100 percent due to rounding.

Data: Automotive News, *Market Data Book*, various issues.

Sources: MVMA, *Facts & Figures*, various issues.

compacts slackened while intermediates and larger cars regained some of their popularity after 1983.

This shift back to intermediates transformed the notion that Japanese cars are compact cars. The actual market shares by class of automobile held by Japanese manufacturers in 1991 were: speciality cars, 53 percent; subcompacts, 43.8 percent; and intermediates, 33.5 percent. Overall, Japanese cars accounted for 29.6 percent of all cars sold in the U.S. The data indicates that, while subcompacts used to be the mainstay of Japanese manufacturers, intermediate and speciality cars have become profitable market segments.[5]

Thus we see that while the two oil crises may have prompted American consumers to opt for smaller cars, this was not the main cause behind the penetration of foreign vehicles into the American market during the latter half of the 1980s. The significant conclusion that we can draw here is that American automakers must now compete with foreign makers in the larger size car market as well (see Table 5.4).

From a demand perspective, the factors that caused auto sales to decline during 1990/91 may be summarized as follows. Auto-related spending may have been affected by the recession in the overall economy, but the degree to which it was affected was similar to or less than that witnessed during the previous recession of 1980/82. The important point to note here is that, of total auto-related spending, the ratio of spending on new cars has clearly declined. Consumers seem to have become more dependent on used cars during the recession of 1990/91, as they were no longer able to keep up with rising new car prices.

Notes

1. See Eric Taub, *Taurus*, New York: Dutton, 1991.
2. This is not to imply that the increase in fleet sales is attributable solely to higher sticker prices. Changes in leasing laws and tax treatment have made fleet sales more attractive. Also, the sale of older fleet cars to buy new ones means a new source of competition for new car dealers and helps to account for the higher ratio of used versus new car purchases.
3. We have so far focused on macro-economic elements, regarding all consumers as a group. While we have been analyzing the "average situation," however, automakers apparently have chosen to focus their attention on a particular segment of the market rather than the whole market. We will take up this subject in Part IV.
4. Using the average price of gas during 1982 through 1984 as the basis for the index, that is, 100, the CPI for gas fell from 108.5 in 1981 to 77.0 in 1986.
5. *Market Data Book*, Detroit: Automotive News, 1992.

   The breakdown of market share shown in Table 5.5 differs somewhat from that in Table 2.4 because vehicles manufactured by transplants are counted as domestically manufactured vehicles. In 1991, however, the sales of Japanese luxury and near-luxury cars exceeded those of European manufacturers. Specific brands of Japanese cars representative of each class are as follows:

Speciality	Mazda Miata, Nissan 300ZX
Subcompact	Honda Civic, Toyota Corolla
Intermediate	Honda Accord, Toyota Camry
Luxury	Acura Legend, Lexus LS400, Infinity Q45
Near-luxury	Lexus ES250/300, Mazda 929, Mitsubishi Diamante

Table 5.5 Segment-by-segment market share in 1991 (percentage)

Brand	Sub-compact	Compact	Inter-mediate	Standard	Near Luxury	Luxury	Total
Domestic	49.7	80.9	64.6	96.2	52.8	63.6	64.9
Japanese	43.8	16.0	33.5	0.0	24.9	19.8	29.6
Other	6.5	3.1	1.9	3.8	22.3	16.6	5.6

Source: Automotive News, *Market Data Book*, 1992.

6

Desperate Efforts at Innovation

When consumers seek to realize their fantasies through automobiles, manufacturers have no choice but to meet their needs. No other product can quite replace the automobile in the ethos of the American public.

Nor can any other transportation form quite match the comfort and convenience that the automobile offers. In this respect, the automotive industry is not directly exposed to competition from any other industry. On the other hand, the public is constantly seeking something new in the marketplace. A new car without an attractive new feature cannot compete with used cars. In the automotive business, a new model may even end up competing with a previous model if the new model fails to offer something new. Asked about competition, Roger Smith once responded, "Our best competition in General Motors right now happens to be a two-year-old Buick."[1]

New-car development is risky. If a new model proves unpopular, the manufacturer can suffer for up to four to seven years after market introduction. Unpopular models also tend to trade at lower prices in the used car market, so that owners of such models are unable to trade in their cars at attractive prices. A failed attempt at new-car development can thus affect a manufacturer's business for more than a decade.

Minimizing risks has become vital in the management of the automotive business. It is not sufficient to merely invest vast sums of money in research and development. Many a new model has failed in the marketplace despite large investments and long periods of preparation.

The structure of the product line, diversification of the market, reinforcement of distribution networks, and streamlining of manufacturing costs are ways in which the risks inherent in product development may be hedged. The more diversified the product line, the smaller the impact of one unpopular model. Having more markets to sell to (such as access to an export market) allows a manufacturer to hedge against the risks—a model that is unpopular in one market may turn out to be popular in another. In some cases, manufacturers can make up for a weakness in product development with their distribution capabilities or price competitiveness. One popular model will not change the ranking in the marketplace, though. Automotive manufacturers are tested by their overall strength.

The overall strength of the Big Three has always been perceived as a threat by the Japanese manufacturers. The Japanese considered the strategies pursued by the Big Three as something that could easily rob them of their position in the American market. Let us herein view the strategy followed by the Big Three in the past by examining how the Japanese perceived them. The Big Three did not, by any means, just sit there and do nothing.

At the beginning of the 1970s, the Big Three were at the height of their prosperity. GM's after-tax profit was higher than the entire market capitalization of Toyota. While Toyota aspired to emulate GM's debt-to-equity ratio, it never dreamed that it would one day surpass GM in terms of absolute profits or in the number of cars it would produce.

The Americans were also way ahead of the Japanese in the area of product development. The Japanese manufacturers were just developing models on a one-off basis, and having a difficult enough time with that. The Big Three, on the other hand, had the big picture in place, with production integrated all the way through to distribution.

Let us focus here on the A-car, which was the mainline product for GM at the time. The A-car was a series of intermediates that shared a common chassis, marketed through four distribution channels: as Chevelle through Chevrolet, as Cutlass through Oldsmobile, as Le Mans through Pontiac, and as Skylark via Buick (renamed Century after 1973). The benefit of marketing an automobile that was basically the same model through four different distribution channels was that it reduced development costs per car. It also offered the benefit of economy of scale on the production line.

On the other hand, the cars needed to be differentiated from one another in a manner appropriate for each distribution channel; otherwise, the models would not sell equally through the four different channels. In other words, if the model projected a luxury image, demand would focus on the Oldsmobile; if cost performance were the chief attraction, it would sell chiefly through Chevrolet.

GM dexterously went about differentiating the products. In terms of design, it installed different types of grills on the cars to project an image that would appeal to each target market segment. The interior was carefully differentiated by class, and the exterior shells were built in differing shapes onto the same chassis by means of innovative welding techniques. This allowed GM to effectively introduce annual model changes. GM went as far as to reshape common parts, such as automotive stampings, to facilitate transportation among the various assembly plants. Product development was conducted based on a thorough understanding of the marketing strategy and the various manufacturing processes.[2]

In contrast, Japanese cars in those days were the sort of cars that were worth only a few lines of mention in Arthur Hailey's 368-page novel:[3]

> "For God's sake, Mr. Trenton! Look at that!"
>
> They examined some of the parts of the Japanese import, the third car they had come to review.
>
> "String and baling wire," Brett pronounced.

The first oil crisis (1973) forced the Big Three to change their product lines. Demand shifted from high-horsepower gas-guzzlers to fuel-efficient compacts. The oil crisis did not strike GM unexpectedly, however; it had already established an Energy Task Force in 1971. Staff of various backgrounds (including manufacturing, marketing, research and finance) were brought together so that the group would have a cross-sectional perspective. The Task Force anticipated an energy crisis in the early part of 1973, which led to the announcement in the autumn of the same year that GM would be downsizing its cars. Immediately after oil prices shot up, a project center was set up to plan the development of smaller cars. GM was thus able to put the Seville and Chevette on the market in 1975.

GM responded to the second oil crisis (1979/80) by introducing the X-car. It invested $40 billion during the first half of the 1980s and planned to raise the proportion of smaller cars that it manufactured. GM needed to manufacture front-wheel drive cars in order to increase the number of smaller, more fuel-efficient cars that it offered. Although these types of cars were smaller on the outside, they offered more room inside the car and were lighter in weight. GM intended to transform itself into the world's largest manufacturer of front-wheel drive cars by raising the proportion of such cars from 28 percent in 1980 to 90 percent by 1985.

A Japanese engineer made the following memorable comment in an interview conducted at the time:

> GM knows what the automobile is all about. We tend to try to create cars that are lopsided on quality, but every part of a GM car from the engine to transmission totally breaks down in the eleventh year, just after the ten-year warranty period expires. They know exactly how/what part functions for how long.

The comment is indicative of the threat that the Japanese manufacturers felt, that is, that if GM were to set its mind on manufacturing compacts, it could easily blow away the popularity that the Japanese had established in the U.S. Obviously, the same notion was applicable to Ford and Chrysler. Everyone thought, at the time, that it was just the beginning of the Japan-U.S. automotive war.

A grand strategy that no other company in the world would have dared to take on was continued. GM acquired Electronic Data Services (EDS) in 1984 to tackle data processing. It went on to acquire Hughes Aircraft in 1985 to involve itself in the aerospace business. In the same year, General Motors Acceptance Corporation (GMAC) was merged to become the largest single financing company in the world. In 1986, GM acquired Lotus, a manufacturer of high-performance cars.

GM was actively involved in other mergers and acquisitions as well. It took a stake in Suzuki of Japan in 1981. With Fanuc of Japan, it established the largest manufacturer of industrial robots in North America in 1982. In 1983, it agreed to set up New United Motor Manufacturing Inc. (NUMMI) with Toyota, which began operations in the following year. It also established Volvo-General Motors Heavy Duty Truck in 1986.[4]

The objective behind these activities was to totally change the notion of "competition."[5] GM did not seek to compete in improved production costs alone; it intended, for example, to load phones and facsimile machines on automobiles, to be operated via communication satellites. It looked to upgrade its ability to offer financing for car purchases. It planned to enhance distribution efficiency by introducing order-entry systems. It also intended to totally automate its plants by introducing industrial robots. In short, it sought to achieve total technological superiority over its competitors through the most advanced forms of high-technology available in the U.S.

However, massive investment in research and development does not guarantee technological innovation. GM's capital expenditure rose sharply from $4 billion in 1983 to $11.6 billion in 1986, but quickly plunged to $5.6 billion in 1988. A number of the investments that had emerged with much fanfare disappeared without achieving their initial objectives.

A good example is the Fiero, a two-seater plastic car that suffered from faulty parts. Also, Buick City, which had introduced the Japanese system of manufacturing, was forced to revert to previous modes of manufacturing. The Saginaw Vanguard, a plant specializing in the production of trans-axles, which was to represent GM's effort of achieving a fully automated plant, was decided to be closed down at the end of 1991. The size of the Saturn Project, GM's compact car program, was curtailed to one-half of its initial size. In short, GM's plan to reverse its competitive position vis-a-vis the Japanese by means of investing $85 billion, which included mergers and acquisitions, was forced to be cut back before being fully implemented.

Meanwhile, NUMMI, a joint venture with Toyota, expanded smoothly. Toyota's manufacturing technology was introduced into a former GM plant in Fremont, California. As a result of dexterous handling of personnel matters, that plant, by no means an ultra-modern facility, was nevertheless able to achieve a level of productivity that was one-and-a-half times that of the average productivity in the U.S. It was able to motivate its work force to work in teams through careful screening of the hirees at the time of recruiting, and through extensive education and training after the work force was hired.

One idea that proved helpful was that managers would go to the production line with a union member in the event of a problem. This approach was known as "joint investigation." In fact, this system reduced the time spent negotiating with the union by 80 percent.[6]

NUMMI owed most of its success to the introduction of systems that involved the handling of inter-personal relationships, which are aspects of the business that are difficult to incorporate into a manual. As a result of this success, many GM personnel were seconded to the plant to learn the Toyota method. However, this approach did not easily fit into other GM plants, because the labor-management relationship at GM was far from cozy (plans were in place to replace a large portion of the blue-collar workers with industrial robots).[7]

Many in the Japanese automotive industry blame the bad performance posted by the Big Three during the 1980s on their being lax about enhancing their competitive capability. In fact, though, the Big Three achieved unprecedented results, even better than they had originally hoped for. Hindsight allows us to say that their strategy was too much of a gamble, and that they strove rather clumsily to achieve leaps in technology almost overnight. Had they won their gamble, though, they would have achieved a quantum leap in technology that no Japanese manufacturer would ever have been able to emulate. The history of the leading American automakers is one of upgrading their competitive capabilities through a series of aggressive investments.

The challenges that the American automotive company undertook during the 1980s had to be undertaken, at some point or other; they were the trump card that it tried to use to compete with the Japanese. The efforts, unfortunately, were in vain. As a result, the Japanese incremental strategy became the mainline approach to manufacturing automobiles.

Figure 6.1 shows the capital investment in the automotive industry (on an actual 1982 value basis) divided by the number of vehicles produced. This index indicates that, on average, some $1,111 was invested per car in 1980 and thereafter. This is one-and-a-half times the average $717 per vehicle invested for twenty years prior to 1980. Although these figures include investments made by Japanese transplants, they are nevertheless an indication that the Big Three made a full-fledged effort to restructure themselves.

The aforementioned capital expenditure did not, however, lead to enhanced production capability. Most of the investment was used for scrapping obsolete plants and rebuilding, introducing metal molds and tools for manufacturing new models, and automating plants as a part of an effort to streamline production. New plants were built, such as Hamtramck (GM) and Atlanta (Ford), but on a net basis, the U.S. automakers' production capability declined.

Figure 6.1 Capital spending per vehicle produced*

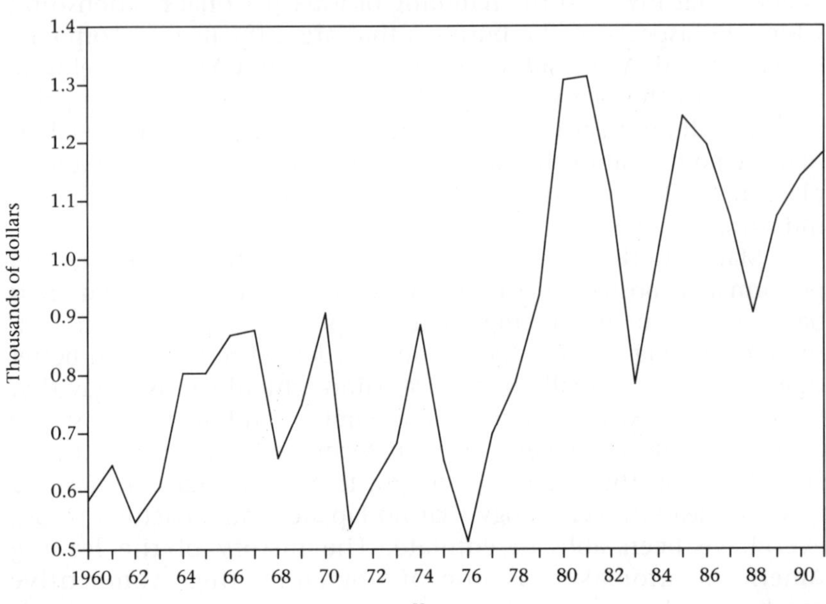

*Capital expenditures on motor vehicles and equipment (SIC 371) in 1982 dollar terms divided by the number of vehicles produced.
Sources: U.S. Department of Commerce; MVMA, *Facts & Figures*, 1992.

Figure 6.2 estimates the production capacity of the U.S. automotive industry (total number of passenger cars and trucks produced divided by the capacity utilization ratio announced by the U.S. Department of Commerce). The production capacity peaked at 14.5 million units in the latter half of the 1970s. In the first half of the 1980s, it fell gradually to 14 million, then further declined to 13.1 million in 1991. Overall, the production capacity of Japanese transplants rose to some 1.9 million units while that of the Big Three declined by 3.4 million units (or by approximately 23 percent from the peak).

A reduction in production capacity is welcome during a recession because it lowers the break-even point of the business. On the other hand, it creates a capacity shortage when demand rises. In fact, during 1985/86 when the economy was expanding, it was said that Ford (which had increased its market share based on the popularity of Taurus and Sable) could have captured an even larger share of the market had its production capacity not been so

Figure 6.2 Production capacity*

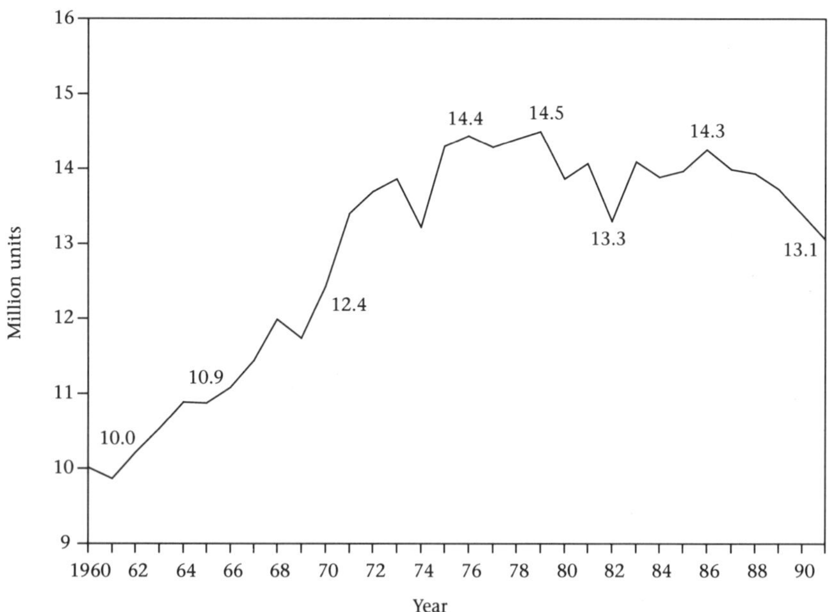

*Number of vehicles produced divided by capacity utilization rate.
Sources: U.S. Department of Commerce; MVMA, *Facts & Figures*, 1992.

constrained. Against the background of declining new auto sales, it is plausible that the Big Three may have given priority to ridding themselves of surplus capacity in order to maintain their financial viability.

Finally, let us briefly examine the number of automobiles produced per worker (inclusive of workers on the assembly line and those in related industries, such as auto parts, and supervisors). Figure 6.3 indicates that productivity changed little during the 1970s and 1980s, but fell significantly in 1991. The average level of productivity from 1968 to 1979 stood at 12.9 units per year per worker, while the average between 1980 and 1990 stood at 12.0. The productivity of production workers, after excluding supervisors, stands at 15.8 units per year per worker for each of the aforementioned two periods.[8]

The annual number of hours worked has remained constant, which implies that the hourly productivity has also remained flat. These facts suggest a negative return on investment when the drastic increase in capital spending after 1983 is taken into account.

Figure 6.3 Labor productivity: Vehicles produced per employee*

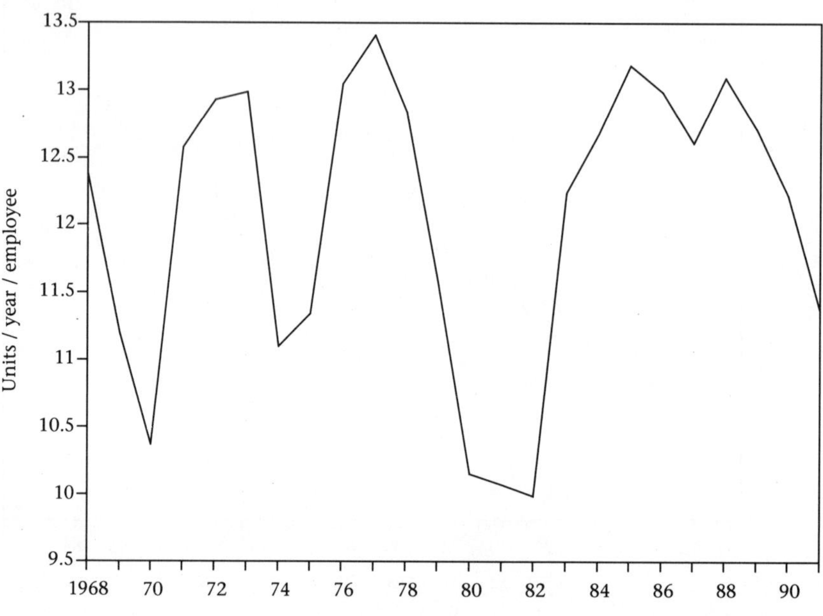

*Annual number of vehicles produced per all employees of motor vehicles and equipment (SIC 371).

Sources: U.S. Department of Commerce; MVMA, *Facts & Figures*, 1992.

The behavior of the Big Three during the 1980s may be viewed as essentially a gamble to recover their competitive capabilities in an environment in which consumers were shying away from purchasing new cars. Imagine how risky it is to make investments when auto purchases are being curtailed! The efforts made by the Big Three might have borne more fruit had they combined their efforts to stimulate the demand for new cars.

In the next part, we will return to the recession of 1990/91 to find out what sort of consumers were able to purchase new cars. This may provide us with a clue as to what might be done to stimulate demand for vehicles.

Notes

1. *Call Me Roger*, p. 296. (Reprinted from *Call Me Roger* by Albert Lee, © 1988. Used with permission of Contemporary Books, Inc., Chicago.)
2. In 1973, A-cars accounted for 23.9 percent of the total number of cars produced by GM (5.5 million). The number of cars manufactured for distribution through the four channels averaged between 230,000 and 400,000 units, which implies that the cars were differentiated very effectively. The cars were assembled at nine plants in six states (California, Georgia, Maryland, Michigan, Missouri, and Texas).
3. Arthur Hailey, *Wheels*, New York: Dell Publishing, 1971, p. 206.
4. Ford pursued a strategy different from that of GM. Ford focused on streamlining its North American Automotive Operations by concentrating efforts on such areas as quality, product, customer-driven philosophy, global strength, its Mazda relationship, technology, and labor negotiations. At the same time, Ford continued to develop new sources of earnings: it acquired First Nationwide Financial Corporation in 1985, U.S. Leasing International in 1987, and Jaguar in 1989.

   Similarly, Chrysler bought Gulfstream Aerospace, E.F. Hutton Credit, and Finance America in 1985; Electro Space System, Lamborghini, and American Motors Corporation (AMC) in 1987; and Thrifty Rent-A-Car and Snappy Rental in 1989. Although Chrysler's financial capability was limited in comparison with its two rivals, the company thought it necessary to prepare for a potential challenge by GM.
5. Maryann Keller, *Rude Awakening*, New York: William Morrow, 1989; Albert Lee, *Call Me Roger*.
6. Toyota Motors, *Nihonsiki Keieini Tuite (On The Japanese Method of Production: The Case of NUMMI*), a memorandum, 1990.

   Before the joint investigation approach was implemented, problems were relayed to the union after they had occurred, which tended to lead to arguments over identification of the problem. By having a union member present at the site to confirm the problem, it was possible to reduce the time spent on confirming the problem, and solutions could be discussed more quickly, which speeded up the whole process greatly.
7. Ford seems to have been able to introduce this type of manufacturing system more smoothly. Ford introduced the Japanese manufacturing system via its tie-up with Mazda, and also reduced surplus capacity by eliminating out-dated facilities. As a result, it was able to match the level of labor productivity posted by Japanese plants. Ford currently boasts the strongest cost-competitiveness among the Big Three companies.

8. The number of employees peaked at 1 million in 1978, of which 781,000 were production workers. This had declined to 800,000 and 609,000, respectively, in 1990. The proportion of workers in administrative and other indirect areas of production rose to 27.1 percent of the total in 1980, but fell to 23.9 percent in 1990. It is expected that the number of administrative staff will continue to be curtailed as the automakers continue their review of white-collar positions.

PART III

Deviating from the Average

7

Immobile Consumers

Between 1988 and 1991, I worked in the research department of Morgan Stanley in Tokyo as a securities analyst. My major concern was the earnings trend of the Japanese automotive companies. Because their profits were heavily dependent on market conditions in the U.S., I also needed to monitor the overall demand and supply situation. My usual approach was to regularly interview a market-analysis expert of a certain Japanese company. Every three months, we would meet to discuss such issues as the macroeconomic situation, demand forecast, trade friction, and the schedule for the introduction of new models.[1]

The first signs of change emerged in the autumn of 1989. A shadow began to be cast over auto sales, which had been buoyant ever since 1983. The market analyst explained the situation very clearly.

"Consumers aged thirty-five and older are losing their appetite for cars. Although our cars are doing all right, I know of other manufacturers that are being affected." Strewn on his desk were documents that gave detailed statistics, broken down by model.

"Take a look at this," he continued. "You can see that the sales of this particular model are down. If sales fall below this line, we will have to start offering sales incentives. Unfortunately, these no longer work as well as they used to, and they may not be effective this time."

"Why are consumers older than thirty-five different from younger people?" I wondered. The specialist seemed to follow my chain of thought.

"Of those Americans who purchase Japanese cars, 46 percent are younger than thirty-five. The ratio is only 29 percent for American cars. It is easier to sell imported cars to younger people because they tend to buy for the sake of quality. Recently, our best customers have been the white-collar, yuppie-types who used to drive European cars before the recession."

"Why does age affect consumer behavior?" I inquired.

"Older people seem to be more sensitive to the state of the economy—retirees, especially. And they are the major buyers of big cars. Sales of these cars have been falling recently, which means that the market share of imported cars will rise again."

I next asked the specialist about the profiles of buyers of different-size cars. He responded by saying that subcompacts compete with used cars and tend to be purchased as entry-level cars by first-time drivers. Housewives and students are the major purchasers of subcompacts, while white-collar workers are the major buyers of compacts and intermediates.

As I looked at one of the graphs on his desk, which indicated a trend in the sales of a particular mainstay car, it was evident that monthly sales were falling gradually vis-a-vis the same month of the preceding year. The expert explained that dealers were competing aggressively for a select group of customers who were still actively buying cars. Both domestic and foreign car dealers were rushing to sell cars to that same market segment.

Incentive programs include a wide variety of actions intended to promote sales. These can be divided into a number of categories: consumer incentives, which offer reduced prices to the consumer; dealer incentives, which offer enhanced sales margins to dealers; lower interest rates on auto loans; and special advertising and promotion budgets. How effective are these?

"That is a matter of the price elasticity of demand. But actually, there are two kinds of price elasticity. One is between 3 and 5, which appears when competitors don't follow suit. In other words, if a manufacturer offers a 10 percent discount, sales go up by between 30 percent and 50 percent. But if competitors also begin offering incentives, the figure falls to 1, or more specifically, to between 0.9 and 1.1. If the manufacturer offers a 10 percent discount, that pushes sales up only by 10 percent. This means that sales value remains virtually flat, but the manufacturer loses because the unit price has been reduced. Manufacturers tend to forget about their rivals when sales fall, though, and are tempted into offering discounts."

"I see. So when the economy is booming, consumers make their choices based on such factors as style, fuel economy, and drivability. But when the economy is weak, price seems to become the major determinant."

"That's right," replied the expert. "But the irony is that it's only the rich that can snap up cars at a discount, because it is the people with surplus cash who have more financial flexibility."

Figure 7.1 traces the share of the Big Three and Japanese automakers in the passenger car market between January 1989 and March 1993. Sales plunged between October and December 1989, the first quarter of the new model year. This is when manufacturers wait to see whether consumers will accept the new models at the prices at which they are offered.

Figure 7.1 Monthly market share of the Big Three and Japanese cars

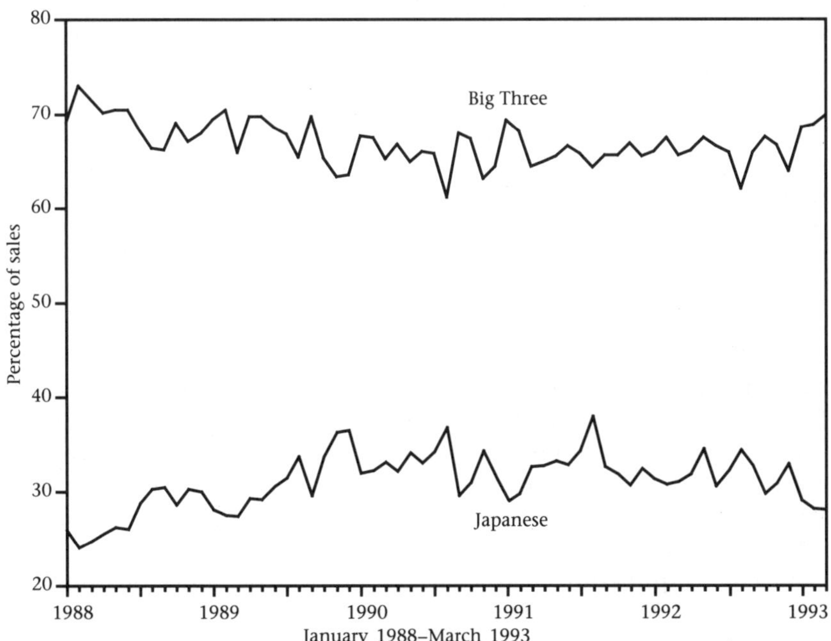

Note: The Big Three includes captive imports.

Sources: *Wards Automotive News*, various issues.

Figure 7.1 indicates that while the Big Three suffered from plunging sales after October 1989, the sales of foreign cars weakened only somewhat. Of note here is the fact that the market share of Japanese cars crept upward during 1990/91. A recession combined with a more conspicuous Japanese presence spells negative sentiment.

In August 1990, Iraqi troops invaded Kuwait, which brought about the Gulf crisis. At the same time, foreign car sales (which includes Japanese cars) began to decline. The prospects of a war in the Middle East robbed the younger generation of their interest in purchasing durable goods. Oil prices shot up, and the possibility of a third oil crisis was in the air.

In January 1991, when fighting began in the Persian Gulf, auto dealers experienced a sharp decline in traffic. Sales activities were toned down as people sat glued to their television screens. Overall January auto sales fell by 25.4 percent from the previous year. Annualized auto sales after seasonal adjustments stood at 7.1 million units, the lowest since January 1982.

I resumed my interview with the specialist in early February.

"It's not just the Gulf War. Sales may not pick up even when the conflict is over. Something fundamental is standing in the way of the rebound we'd like to see."

I had asked how long it would take for auto sales to recover after the fighting had ceased. The reply was totally contrary to my expectation that pent-up demand would resurface once the war was over, and that sales would pick up. It seemed that I was being optimistic.

"Actually, 1991 shouldn't be a bad year, at least in view of the cyclical nature of the business. New cars sold well during 1985 and 1986, and 1991 would be the year in which those purchasers will be looking to replace their cars, because auto loans have been extended to five to six years. Most loans used to run for less than four years. The term of the loans has been extended to promote sales, however, and most outstanding loans these days are five- to six-year loans.[2] Most people don't buy new cars until their loans are paid off, so people might start replacing their cars sometime in the latter half of 1991."

The implication was that, as far as the auto business is concerned, 1991 would be a year for replacement purchases: good news for securities analysts in charge of forecasting future demand and corporate earnings. Auto sales have remained strong regardless of the slow-down in the overall economy. The above paints an appealing scenario for the outlook of the automotive industry.

"But, you know," he continued, "something is funny. Only certain types of customers seem to be buying, and total business is still slow."

Funny, indeed. Something dramatic must be happening. I leaned forward to listen closely.

"In 1990, domestic cars were discounted by about $1,000 per car, which is about 7 percent off the sticker price. Japanese cars were discounted by 5 percent, or $800. We've rarely seen such price reduction previously. Yet, this hasn't worked well. It's almost like abusing morphine. Incentives after incentives have been offered to no avail, and it looks as though we've finally reached a point of saturation. Why, do you think?"

"Don't ask me, I should be asking the question," I thought. What did occur to me, however, was the theory of price elasticity. If economic theory doesn't work, there must be something else. Have consumers become anti-Japanese and thus started avoiding Japanese cars?

"What it boils down to, in the end, is that auto prices have been hiked too many times in the past. We expect 1991 auto sales to be between 8.5 and 9 million units, but had there been no price increases since 1980, the figure may have been more like 12 million."

Auto sales peaked at 11.46 million units in 1986. Extrapolating from there, 12 million sounds like a plausible number. The automotive company for which the specialist worked is known to use a sophisticated demand forecast model that employs six factors as explanatory variables: family income, interest rates, number of models introduced or remodeled, inflation rate, new car prices, and used car prices. Their estimates have so far been incredibly accurate.

"What really surprised us was that, if high auto prices were the problem, sales should have recovered when larger discounts were offered. But there was no reaction. Discounts don't work after a succession of price increases."

It seemed that the effects of price increases and discounts were not necessarily symmetrical.

"A new car costs more than $15,000 these days. To afford this kind of a price, an annual family income of at least $48,000 is required. The average, however, is $34,000, which means that average families can't afford to buy a new car."[3]

I was getting lost. "Give me the total picture," I requested.

"The figure $48,000 is roughly 40 percent higher than the average family income. What proportion of families, do you think, earn this type of income? Only one-third; only the top third of families in the U.S. can afford to buy a new car, that's what this implies."

"So, it seems that we must not assume that all consumers are potential buyers of new cars," I said. "New car buyers are now limited to the more affluent."

"Actually, our analysis indicates that active buyers of new cars own more than one car. Such households tend to own two or three cars.[4] We speak of 'households,' but when we look at the family composition of such customers, they usually consist of only one or two members.[5] These are the so-called DINKS: Double (or high) Income, No Kids. They buy new cars and supply used cars. That's the current situation in a nutshell."[6]

The rise in the number of cars in use indicates that more new cars are being bought than older cars being scrapped. This implies that a select segment of the market purchases new cars, while others drive used cars for a longer period of time.

"As you know, even high-tech giants like IBM and Boeing have begun to lay off middle managers. Working for a blue-chip outfit no longer implies job security. This is known as the white-collar recession."[7]

Indeed, it was not just the automotive industry that was feeling the pinch. IBM has announced losses. The computer/electronics industry, long believed to be an ever-expanding industry, was no longer recession-proof. And the defense industry had to cope with the post-Cold War era.

"People won't buy cars, even if offered at a discount, when they feel insecure about their jobs. Besides, it's usually their second or third car, so there's no immediate need to buy. You might ask, then, why not try selling cars to the lower-income earners? A 10 percent discount isn't going to work with this group, and it's risky to offer them financing: it's not an issue of what rate, it's an issue of credit. They are already up to their necks in loans. Car prices kept being raised, and we've gone just about as far as we can go."

The picture was becoming a bit clearer. Dealers had been catering to high-income earners, but have now reached a point where either income growth was no longer keeping up with the rise in car prices, or a smaller number of people were earning high incomes. Meanwhile, dealers were unable to attract other types of customers. A substantial discount would have to be offered to stimulate interest among the lower-income earners.

On the other hand, cars do come in a variety of sizes, from subcompact to full size. In the 1992 model year alone, 555 different new models were offered.[8] Even if the average price of a car has risen, the automakers should be able to sell cheaper models.

"What cheaper models? There aren't that many manufacturers of cheaper cars left," responded the analyst when I inquired about this point. "Japanese cars are now more expensive than American ones, and the supply of Korean and Yugoslavian cars hasn't increased. You're probably correct in saying that cheaper cars will sell better. Saturn should do well. But if automakers concentrate on cheaper lines, the business will generate substantial losses."[9]

"Then, there's no way out?" I asked.

"Actually, what is now supporting the business is sales of so-called program cars. This is basically selling fleet cars for rental and other purposes under a special contract: there is a buy-back condition attached. Manufacturers will buy back the used car in three or six months. This is the ultimate discount, up to 20 to 30 percent. This type of business now accounts for some 20 percent of total sales, equivalent to 1.5 million cars per year.[10] This is the only area that is growing over previous years. Non-program car sales are down substantially.[11] The drawback is that these program cars will be resold at a further 20 to 30 percent discount in the used car market, pushing down the price of second-hand cars and creating yet another reason why the consumers will shy away from buying new cars. Lower-income earners would rather look for a car in the used car market."

So, although demand for fleet cars is rising, the business cannot generate much profit if prices are discounted by as much as 20 to 30 percent. Automotive marketing strategy seems to be facing a dead end.

The above scenario suggests that a major change is about to occur in the automotive industry. The industry has always been regarded as a leading indicator of business conditions. It is sensitive to falling interest rates and among the first sectors of the economy to recover. That, at least, was the case until recently.

In the future, assuming that auto sales depend on a resurgence in white-collar jobs, the industry could become a laggard. Why? Because companies do not increase white-collar jobs until their business improves. And for business to improve, production must first pick up, which implies that blue-collar jobs will increase before white-collar jobs. Meanwhile, however, a fall in blue-collar unemployment will not help new auto sales, which means automotive sales will pick up not at the beginning of an economic recovery, but later on.

If the recovery in auto sales lags behind the general economy by a substantial margin, the industry may see interest rates rise before its business recovers, which could pose a major problem for equity investments. If interest rates rise, stock prices face a higher probability of falling. This suggests that the absolute levels of automotive stocks may fail to rise regardless of recovering auto sales.

In a nutshell, the conclusion to be drawn from the interviews is that it will be some time before the automotive industry can recover. Unemployment statistics will be crucial. Auto sales failed to rebound during the summer of 1991. We see no improvement in Massachusetts, Florida, or California, where numerous white-collar jobs exist. Auto sales in the Midwest, home to many manufacturing plants, however, did relatively well. The prophecy thus showed signs of coming true.

At a GM analysts' meeting held in early August 1991, the company officially announced its concerns over the possibility of a prolonged recession. The introduction of the 1992 models in October 1991 failed to stimulate sales. Even a succession of interest rate reductions did not help. The ultimate key is the relationship between income growth and the rise in auto prices.

In the next chapter, we will wrestle once again with some statistics to review automakers' pricing policies. In Chapter 9, we will then examine the crucial subject of income distribution.

Notes

1. The style of description employed for much of this chapter and Chapter 17 is that of an imaginary dialogue between a U.S. market researcher in a Japanese automotive company and the author of this book. I have taken this approach for three reasons: it allows readers outside Japan to eavesdrop on a typical discussion among the Japanese experts, it offers an opportunity to understand the depth of Japanese research on the U.S. economy, and it enables all basic perspectives to be displayed in a comprehensive manner. In fact, during 1990–91, Japanese automakers were able to accurately estimate U.S. auto sales, and their methodology was well coordinated, facilitated by their more than ten years of accumulated data. This has had a major impact on my writing this book.
2. In four years, from 1986 to 1991, the term of auto loans was extended. Although 49 percent of contractors used financing of 37–48 months and 27 percent contracted for 49–60 months in 1986, by 1990, 46 percent of them had started using contracts of 49–60 months and just 32 percent depended on 37–48 months. See MVMA, *Facts & Figures*, 1988, p. 64, and 1991, p. 61.

3. "1990 buyers of new cars," *Newsweek;* quoted in MVMA, *Facts & Figures,* 1991, p. 46.

   As used here, annual income means "family earnings." The Census Bureau differentiation between "family" and "household" is as follows (source: *The Annual Report of The Council of Economic Advisers,* 1992, p. 117):

   Family: a group of two or more people related by birth, marriage, or adoption who live together.

   Household: all related family members and all unrelated people living in a given housing unit.

   The basic difference, then, is that a "household" includes single persons while "family" does not. Singles tend to be younger, and they earned a median household income of about $29,943, approximately 15 percent less than the median family income of $35,353 in 1990. The number of these respective groups is 94,312,000 and 66,322,000 (U.S. Department of Commerce, Bureau of the Census, *Statistical Abstract of the United States,* 1992, Tables 697 and 702).
4. From 1973 to 1988, the diffusion of passenger cars increased from 83 to 88 percent. During that same period, the rate of multi-car ownership increased from 39 to 51 percent, and the ratio of families owning three or more cars skyrocketed from 2 to 14 percent.
5. This situation, of course, is not just applicable to car buyers. On a nationwide basis, the total number of households in 1990 was 94.32 million, of which 23.59 million were single-person households and 30.18 million were two-person households. One- and two-person households thus account for more than half of the total households (*Statistical Abstract of the United States,* 1992, Table 697).
6. On a unit basis, the proportion of used cars among passenger cars sold has risen steadily, from 60.0 percent in 1989 to 61.6 percent in 1990 and 65.0 percent in 1991.
7. The unemployment rate by occupation indicates that unemployment among white-collar workers has been increasing since 1988. In 1991, the unemployment rate for white-collar workers was about 3 percent while that for blue-collar workers and workers in sales, services, and support was 9 and 6 percent, respectively. Although 3 percent is below the 3.5 percent posted when unemployment peaked in 1983, the very fact that the proportion of unemployed white-collar workers has risen seems to have led to the phrase, "white-collar recession." Note especially Chart 2–11 in *The Annual Report of The Council of Economic Advisers,* U.S. Government Printing Office, February 1992.
8. *Automotive News,* May 4, 1992, p. 14.
9. The Saturn was first marketed in 1991 at a basic price of $8,270 for a four-door sports sedan and $12,050 for a two-door sports coupe. These prices were lower than initially expected. Sales have been strong, with

74,498 units being sold in 1991. The average number of units sold per dealer outlet was 776 in 1991, making it the best among competitors. Also, the Saturn was ranked third on J.D. Power's Customer Satisfaction Index (after the Toyota Lexus and the Nissan Infinity). On the down side, GM spent 3 billion dollars on developing the Saturn. According to Maryann Keller, "the $3 billion price tag for this 'import fighter' makes the victory empty" (*Rude Awakening*, p. 267).

10. *Automotive News*, October 14, 1991, p. 43.
11. Fleet sales in 1991 increased by 12.4 percent over the previous year, while retail sales fell by 20 percent.

8

Limited Availability

The most important information gleaned from the dialogue with the specialist described in the previous chapter is that car prices have been rising so rapidly that an entire class of consumers is now unable to afford new cars. In this chapter, the trend in car prices will be compared with family income and general price increases.

The list price of a car is an easy concept to understand: it is marked on the price tag at the dealer's shop and cited in many car magazines. However, an analysis of car prices over a long period of time requires a careful procedure. The analytical steps employed here should be explained briefly.

There were 555 car models offered in the 1992 model year. On top of this, there are numerous combinations of available options, such as air conditioning, brakes, paint, roof, and wheels. So what, then, is the best estimate of price? Is "average new car price" the most representative one? The difference between the actual price of a model and the average price must be carefully considered. Also, the prices of the same model for different model years must be carefully adjusted for precise comparison.

Let us consider an example of these differences using a hypothetical case. Suppose there were only two models, Expensive and Cheap, and that their sales records in 1991 and 1992 were tabulated as follows:

	1991			1992		
	Sales units	Average price	Sales value	Sales units	Average price	Sales value
Expensive	10	$100	$1,000	20	$110	$2,200
Cheap	90	$10	$900	80	$10	$800
Overall	100	$19	$1,900	100	$30	$3,000

In this example, the price of the Expensive model rose just 10 percent between 1991 and 1992, and the price of the Cheap model remained the same. Yet the average price increased by an astonishing 57 percent, from $19 to $30.

The growth rate of the average price masks changes in the model mix. To pick up the net change of each model price, therefore, the sales units are assumed to be the same in both years. If we take 1991 as the base year, the average of the price increases weighted by the sales units would show around 5.2 percent increase during the period.

However, the 1991 Expensive model might not be the same in quality as that of the 1992 model. An Anti-lock Brake System (ABS), which was optional in the 1991 model, may have become standard equipment in 1992. In that case, the value of the basic model of 1991 plus the cost of the ABS should be compared with the price of the 1992 model. In the calculation of Consumer Price Index (CPI), such adjustment is made strictly.

Regarding actual automotive prices, the CPI of cars has increased less than the growth rate of median family income, yet the average price has surged significantly, as shown in Table 8.1. A straightforward conclusion that could be drawn from this is that manufacturers have not raised the price of cars; rather, consumers have chosen upgraded models. So the question then becomes whether consumers have voluntarily shifted their preferences to more expensive cars, or whether they were compelled to do so by the manufacturers and dealers. This issue will be discussed later using Table 8.2.

A simple examination of the price trend of a certain class of model, unadjusted for any quality change, suggests the availability of the model. Again, in the above example, even if the net price did not change, some consumers in 1992 would have become unable to buy the same model as the 1991 Expensive without ABS. This type of pricing enables the manufacturers to select the purchasers of their product.

Table 8.1 Consumer expenditure per new car, new car prices, and annual median income (in dollars)

Year	Consumer expenditure per new car (1)			Estimated average new car price for a 1967 "comparable car" (2)	Annual median income (3)	Producer price index (1982 = 100) (4)
	Average	Domestic	Import			
1970	3,542	3,708	2,648	3,601	9,867	38.0
1975	4,950	5,084	4,384	4,686	13,718	56.0
1980	7,574	7,609	7,482	6,863	21,023	91.7
1981	8,910	8,912	8,896	7,700	22,388	98.7
1982	9,890	9,865	9,957	8,078	23,433	100.0
1983	10,640	10,559	10,873	8,387	24,580	101.2
1984	11,450	11,172	12,354	8,685	26,104	104.1
1985	12,022	11,733	12,875	8,984	27,144	103.3
1986	12,894	12,526	13,815	9,395	28,236	102.2
1987	13,657	13,239	14,602	9,743	29,744	105.3
1988	14,468	14,029	15,537	9,995	30,992	113.2
1989	15,292	14,957	16,127	10,248	32,448	118.1
1990	16,012	15,641	17,010	10,581	33,969	118.7
1991	16,700	16,152	18,198	11,152	34,775	118.1
1975/70	139.8	137.1	165.6	130.1	139.0	147.4
1980/75	153.0	149.7	170.7	146.5	153.3	163.8
1985/80	158.7	154.2	172.1	130.9	129.1	112.6
1990/85	133.2	133.3	132.1	117.8	125.1	114.9

(1) U.S. Department of Commerce, Bureau of Economic Analysis, "Average transaction price per new car."

(2) 1967 "Average consumer expenditures per new car" plus the value of added safety and emissions equipment as determined by the U.S. Bureau of Labor Statistics, all inflated to current dollars, using the U.S. Bureau of Labor Statistics, "New car consumer price index—All urban consumers."

(3) U.S. Department of Labor, Bureau of Labor Statistics, "Median family earnings."

(4) U.S. Department of Labor, Bureau of Labor Statistics, "Producer price index, intermediate materials, supplies and components, materials and components, for manufacturing."

Sources (1)–(3): MVMA, *Facts & Figures*, various issues.

Some readers may feel that the retail price is decided not only by the manufacturer, but by the dealers as well. While this is true in the short-term, in the long-run it is the manufacturer's intent that is the crucial determinant. If we compare the average shipment price of cars with the average retail price of cars between 1977 and 1991, the markup is relatively stable (at around 16 percent) except for years in which the supply-demand situation was extraordinarily tight or easy. Therefore, we can concentrate on the price comparison as follows.

In Table 8.1, we can see first how the average consumer expenditure per new car (equivalent to the weighted average price in the above example) has deviated from the estimated average new car price for a 1967 "comparable car." This reflects exactly the trend of the New Car Consumer Price Index—All Urban Consumers. The average expenditure in 1991 was $16,700, but the 1967 "comparable car" price was just $11,152. The difference has amounted to almost 50 percent in the last twenty-five years.

More specifically, in the five-year increments shown at the bottom of the table, the average expenditure has increased faster than the new-car CPI. Moreover, the CPI of all items increased more sharply than the new-car CPI. The general price index rose 139 percent from 1970 to 1975, 153 percent from 1975 to 1980, 131 percent from 1980 to 1985, and 122 percent from 1985 to 1990.

In short, although the prices of new cars have not increased so rapidly (after adjusting for the quality change), the average expenditure has surged even higher than the general inflation rate. This means either that consumers have upgraded their tastes toward more expensive cars, or that the availability of new cars has been limited to a richer class of consumers.

The rate of increase of average expenditure was almost the same as the Annual Median Income of families in the 1970s. But in the 1980s, and especially in the first half of the decade, the former outpaced the latter significantly. The auto manufacturers seemed not to consider consumers' financial positions.

The same pattern of comparison is applicable to the Producer Price Index of intermediate materials, supplies, and components, which is taken as the proxy for the input price index of the automotive industry. There was parallel movement between the two indices in the 1970s, but after the first half of the 1980s the average expenditure began to neglect the shift in the cost curve.

All-in-all, the average expenditure moved free of any restraints of consumer income and production cost in the 1980s. In general, these two factors will decide the market value of new cars in a competitive market.

What kind of mechanism has been functioning in the pricing of new cars? Before proceeding to this topic, we need to check whether the rise in the average expenditure is due to a shift in consumer preference or to the manufacturers' pricing strategy. Of course, even if consumers could buy cheaper cars but choose instead to buy more expensive ones, the case could be thought of as a matter of preference. If the availability of cheap cars were to be limited, however, the pricing policy would be the problem. The competitive situation in the marketplace in the 1980s, then, will be discussed by looking at the pricing of Japanese imports.

It is indeed necessary to check whether consumers spontaneously upgrade the car that they purchase, or whether automakers have simply refrained from promoting cheaper cars. An examination of the cheapest model available for each model year (Table 8.2) will serve to indicate the availability of new cars to consumers who can afford only the cheapest new model.

Table 8.2 Trend of lowest new car price

Brand	1970	1975	1980	1985	1990
GM*	2,024	2,799	3,782	5,760	8,002
	(100)	(138)	(187)	(284)	(395)
Ford*	1,885	2,919	3,781	5,928	7,742
	(100)	(155)	(201)	(314)	(410)
Chrysler*	2,157	3,269	4,295	6,174	7,358
	(100)	(152)	(199)	(286)	(341)
Toyota	1,726	–	4,098	5,348	6,753
	(100)	–	(237)	(310)	(391)
Nissan	1,935	–	3,794	5,499	7,299
	(100)	–	(196)	(284)	(377)

*Excluding captive imports.

Sources: *Ward's Automotive Yearbook*, various issues.

It is difficult to directly compare the various pricing strategies pursued by different automakers; individual strategies depend on whether the year is one for model changes and on the market share that they were able to capture during the previous year. However, Table 8.2 indicates that the price of the cheapest model has risen considerably regardless of the manufacturer when compared to the prices of comparable models of 1967. If the price of the cheapest model had risen at the same rate as income growth between 1970 and 1990, the index should stand at 344 in 1990. Chrysler's 341 is the only index that is below this figure, but note that its 1970 price level was the highest. This examination thus sheds some light on the limited choices for low-end consumers.

Let us now consider the degree of competition in the early 1980s. Note that, as shown in Table 8.3, the average consumer expenditure per domestic car was overtaken by that for foreign cars after 1984. The prices of foreign cars rose by 143.3 percent between 1980 and 1991, while domestic car prices increased only by 112.2 percent. Why did this different pricing occur?

In 1979, oil-consuming countries were faced with a second oil crisis, and regulations on fuel economy and emission were

Table 8.3 Car price comparison: Domestic vs. import

Year	Consumer expenditure (1)		Japanese import (2)				Yen rate (3) (yen/$)	Japanese export price index (4)
	per domestic new car ($)	per import new car ($)	Quota* (000)	Unit (000)	Unit value ($ mil.)	Dollar value ($)		(1980 = 100)
1980	7,609	7,482	0	1,992	8,229	4,132	226.74	100.0
1981	8,912	8,896	1,680	1,912	9,491	4,965	220.54	104.1
1982	9,865	9,957	1,680	1,823	9,627	5,281	249.08	111.6
1983	10,559	10,873	1,680	2,112	11,064	5,239	237.51	110.3
1984	11,172	12,354	1,850	2,692	13,129	4,878	237.52	113.3
1985	11,733	12,875	2,300	2,527	16,176	6,400	238.54	113.2
1986	12,526	13,815	2,300	2,619	21,061	8,042	168.51	100.2
1987	13,239	14,602	2,300	2,418	21,305	8,813	144.62	97.6
1988	14,029	15,537	2,300	2,123	19,878	9,363	128.24	93.9
1989	14,957	16,127	2,300	2,052	20,152	9,823	137.98	99.7
1990	15,641	17,010	2,300	1,868	19,500	10,440	144.82	103.8
1991	16,152	18,198	2,300	1,789	20,631	11,531	134.54	101.7
1992	-	-	1,650	1,637	20,069	12,259	126.65	102.3
1991/80	212.2	243.3	-	89.8	250.7	279.1	59.3	-

*The voluntary export restraint quota applies to the fiscal year ending March of the following year. Also, the Japanese export classification does not include some vans and wagons that are counted as passenger cars in the U.S. import statistics.

Sources: (1) MVMA, *Facts & Figures*, 1992.
(2) U.S. International Trade Commission, *The U.S. Automobile Industry Monthly Report on Selected Economic Indicators*, various issues.
(3), (4) The Bank of Japan, *Price Indices Annual*, various issues.

tightened. Demand shifted to smaller cars, and auto sales declined substantially. Only imported cars fared well.

In 1980, dumping allegations were filed against Japanese cars. The allegations were later denied, but in April 1981 the Japanese government set a voluntary quota on car exports. Imported cars were selling well in the U.S., but the supply was squeezed; consequently, an imbalance occurred between supply and demand. The market for Japanese cars became a sellers' market. Japanese cars disappeared from dealer showrooms, and buyers were forced to wait from three to six months for delivery.

The appropriate level of a dealer's inventory is said to be sixty-days' worth of sales. In the case of Honda and Toyota, this was down to about ten days. Japanese cars sold at a premium. The price of Japanese cars, which had previously represented fierce price competition for the American auto manufacturers, rose sharply.

The years 1980–82 saw an unprecedented recession. Under normal circumstances, retail prices would have fallen during this time, as the supply exceeded demand. However, prices of domestic cars went up, as if pulled up by prices for imported cars.

Table 8.3 also gives a breakdown of the number of Japanese cars imported (voluntary quota and actual), the value, (average) unit value, the dollar-yen exchange rate, and the Japanese export price index (based on the yen).

The price of cars imported from Japan had been monitored closely since the latter half of the 1970s, due in part to the investigations conducted by the Department of Commerce in conjunction with the dumping allegations. Japanese manufacturers were not allowed to set export prices substantially lower than the prices in Japan, which implied that the dollar-denominated export prices had to be hiked when the yen strengthened against the dollar.

Meanwhile, the Internal Revenue Service in the U.S. was investigating transfer pricing. If the dollar-denominated export prices were raised by a larger margin than the sticker price offered in the U.S., the importers would see their profit margins narrow, which meant income tax that would have been payable in the U.S. would be paid in Japan. To avoid this, the manufacturers had to link export prices to sticker prices to prove that transfer prices were not being manipulated intentionally.

Japanese manufacturers ended up mechanically linking the export prices with sticker prices on their cars during the course of the 1980s to avoid complaints on either front. Owing to the voluntary restrictions on car exports from Japan, the demand for Japanese cars exceeded supply in the first half of the 1980s.

Japanese car prices rose sharply in the U.S., and the price differential between the prices in the U.S. and those in Japan widened to as much as 45 percent.[1] It was estimated that pre-tax profitability on car exports to the U.S. rose to as high as 30 percent.[2]

Did domestic car prices rise as a result of higher Japanese car prices? Or had the Big Three pursued a strategy of driving prices higher? Table 8.3 again sheds some light on the situation.

In summarizing the pricing policy of the U.S. automotive companies, the following quote is very helpful:[3]

> As a result, new cars have become significantly more expensive and less affordable commodities. Two other observations are relevant in this regard. First, the rapid rise in new car prices *cannot* be attributed to costs of government regulation of automobile safety and pollution. Second, auto companies have *raised* their prices at an accelerating rate, even as they boasted about efficiency gains and productivity improvements. Here again, import quotas and the elimination of foreign competition appear to have played a key role.

The Big Three are thus deemed to have pursued a policy of sustained price increases regardless of lower manufacturing costs and higher profitability. This becomes even more evident if we focus on the events of the latter half of the 1980s. After the Plaza Agreement of September 1985, the dollar plunged from 240 yen to 120 yen against the dollar, and the resultant effect was sharp rises in both the export and retail prices of cars exported out of Japan. The Export Price Index of Japan was virtually flat after 1985, which means that the dollar-term export prices rose in parallel with the yen's appreciation. The market for passenger cars recovered in 1985–86, and the Big Three posted record profits during this period. Further price increases were hardly necessary, but the Big Three seem to have opted for higher prices rather than maintaining prices to increase their market share.

It is a fact that the automakers were then operating at full capacity. This implies that they were unable to produce more cars, which they could have sold at a discount to capture a further share of the market. In Ford's annual report of 1986 we find the following comment:[4]

> By the end of 1986, Ford had produced at capacity for 29 consecutive months. At high industry sales volumes, capacity restraint was a major reason for the 1986 decline in our car market share.

U.S. automakers were faced with a strong need to finance their capital investments, and larger profits would have helped greatly. On the other hand, if they were to opt for higher prices, it became more likely that they would suffer from a steeper slump in sales in the event of an economic downturn. Auto sales might not have declined as much as in 1990–91 had the automakers restrained price acceleration to within the rate of inflation, as the Japanese automotive expert had forecast in the previous chapter.

In closing, why did the Big Three opt for sustained price increases? They must have believed that the price hikes would not affect sales. To reach a definitive conclusion, however, we must examine the distribution of income in the U.S.

Notes

1. Organization for Economic Cooperation and Development (OECD), *The Cost of Restricting Imports, The Automotive Industry*, Paris: OECD, 1987.
2. Table 8.3 indicates that the average unit price of imported Japanese cars declined between 1983 and 1984. This is believed to have been due to the increased proportion of compact cars exported to the U.S. Further research indicates that the average expenditure on a Japanese car was $6,709 in 1980, $7,292 in 1981, $7,538 in 1982, $8,317 in 1983 and $9,229 in 1984. The unit prices are inclusive of dealer's premium, however, and do not completely reflect the pricing policy of the Japanese automakers (Charles Collyns and Steven Dunaway, "The cost of trade restraints. The case of Japanese automobile exports to the United States," *International Monetary Fund Staff Papers*, Vol. 34, No. 1, 1987, pp. 150–79).

   The report concludes that the "voluntary restrictions" cost the American consumers $16.75 billion between 1981 and 1984, and between 4 to 7.5 million jobs in the automotive industry.

   For a detailed examination of the upgrading of Japanese cars exported to the U.S., see Robert C. Feenstra, "Voluntary export restraint in U.S. autos, 1980–81: Quality, employment, and welfare effects," in Robert E. Baldwin and Anne O. Krueger, eds., *The Structure and Evolution of Recent U.S. Trade Policy*, Chicago: University of Chicago Press, 1984.
3. Walter Adams, ed., *The Structure of American Industry*, 7th Edition, New York: Macmillan Publishing, 1986, p. 152. (Reprinted with the permission of Macmillan Publishing Company from *The Structure of American Industry*, Seventh Edition, by Walter Adams. Copyright© 1986 by Walter Adams.)
4. Ford, *Annual Report 1986*, p. 2.

9

Growing Income Dispersion

This chapter presents several tables and figures that illustrate the degree of income dispersion in the U.S. The implications it holds for the automotive industry will be discussed in Part IV.

Figure 9.1 illustrates the widening income dispersion in the U.S. The Gini Coefficient (an index of income concentration based on Gini's law) rises after bottoming out in the latter half of the 1960s. *The Annual Report of the Council of Economic Advisers 1992* has the following to say on the subject of a widening income differential:[1]

> Since the mid-1960s and in particular since the early 1980s, income growth has occurred in all quintiles and the distribution of annual money income has become more dispersed in the United States.... Families and households display significant mobility across income classes. The distribution of long-term income is more equal than the distribution of annual income.

The government thus acknowledges that the degree of dispersion of income distribution has increased. Since personal income fluctuates with age, however, it claims that the differences in lifelong income have not widened as much as indicated in Figure 9.1.

Figure 9.1 Gini ratios of family income

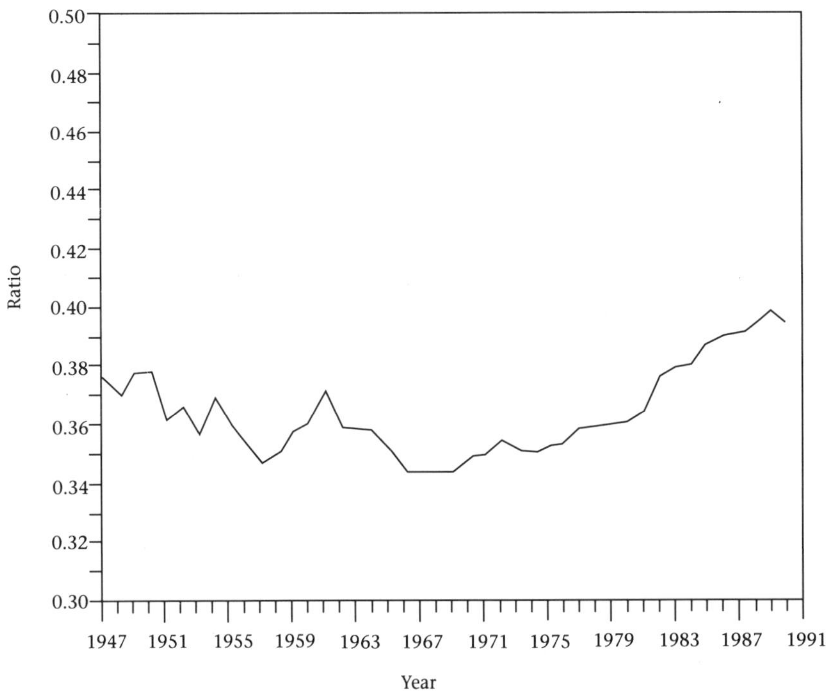

Data: Department of Commerce and Department of Labor.
Source: *Economic Report of the President*, February 1992, p. 123.

Figure 9.2 examines which income classes have expanded in terms of the number of families. After 1967, families belonging to the highest income bracket (those earning an annual income of $50,000 or more, in 1990 constant dollars) rose. Meanwhile, there was only a slight reduction in the number of families in the low-income bracket (with annual incomes of below $15,000). The number of families in the middle income bracket, however, fell sharply.

The proportion of families in the highest income bracket is most affected by cyclical fluctuations in the economy. As Figure 9.2 indicates, the proportion declined during the recession after the first oil crisis in 1973, during the 1980–82 recession, and again in 1990. Let us recall, as suggested in the previous chapter, that buyers of new cars in 1990 needed to earn a median family income of $48,000. Fluctuations in the number of families in this "high-income" group, therefore, must have substantial impact on the sales of new cars.

Figure 9.2 Distribution of families by income class*

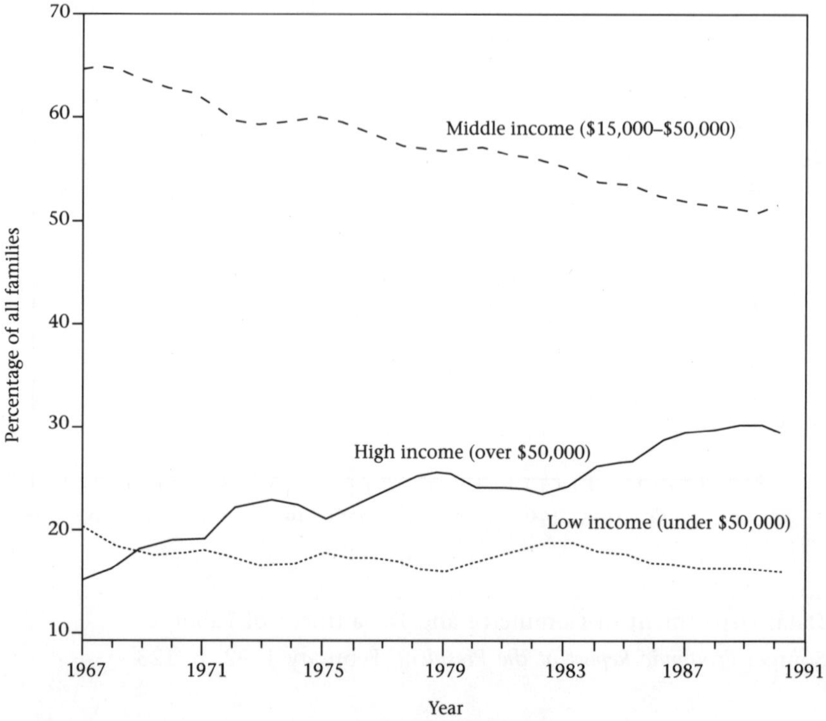

*All income is in 1990 dollars: the corrected Consumer Price Index for all urban consumers (CPI-U-X1) is used as the deflator.

Data: Department of Commerce and Department of Labor.

Source: *Economic Report of the President,* February 1992, p. 121.

Tables 9.1 and 9.2 are based on statistics that categorize families into five equivalent groups (quintiles, or divisions of 20 percent based on family income), beginning with those earning lower incomes.[2] Table 9.1 shows the highest income earned by families grouped in the lowest income through the second highest income classes. Since there is little point in showing the top income earned in the highest income class (which would be the income of a single, that is, the nation's highest earning, family), this group (the highest income quintile) is replaced with the group with family income that is 5 percent below the top.

Table 9.1 Family income at selected positions

Year	Median income ($)	Upper limit of each fifth				
		Lowest ($)	Second ($)	Third ($)	Fourth ($)	Top 5% ($)
1970	9,867	5,100	8,320	11,299	15,531	24,250
1971	10,285	5,211	8,628	11,826	16,218	25,325
1972	11,116	5,612	9,300	12,855	17,760	27,836
1973	12,051	6,081	10,034	14,916	19,253	30,015
1974	12,902	6,628	10,694	15,015	20,690	32,199
1975	13,719	6,914	11,465	16,000	22,037	34,144
1976	14,958	7,441	12,400	17,300	23,923	37,047
1977	16,009	7,903	13,273	18,800	26,000	40,493
1978	17,640	8,700	14,700	20,600	28,600	44,900
1979	19,587	9,800	16,200	23,000	31,600	50,300
1980	21,023	10,286	17,390	24,630	34,534	54,046
1981	23,288	10,918	18,552	26,528	37,457	58,554
1982	23,433	11,200	19,354	27,750	39,992	64,000
1983	24,580	11,629	20,060	29,204	41,824	67,326
1984	26,433	12,489	21,709	31,500	45,300	73,230
1985	27,735	13,192	22,725	33,040	48,000	77,706
1986	29,458	13,866	24,020	35,015	50,370	82,273
1987	30,970	14,450	25,100	36,600	52,190	86,300
1988	32,191	N.A.	N.A.	N.A.	N.A.	N.A.
1989	34,213	16,003	28,000	40,800	59,550	98,963
1990	35,353	16,846	29,044	42,040	61,490	102,358
1991	35,939	17,000	29,111	43,000	62,991	102,824
1991/70	364.2	333.3	349.9	380.6	405.6	424.0

Sources: *Statistical Abstract of the United States,* various issues.

Table 9.1, which covers the period between 1970 and 1990, indicates that income has grown at a faster pace for classes with higher incomes. In other words, the rate of income growth in the highest income class was the highest, while the rate for the lowest income class was the lowest.[3]

Let us turn now to Table 9.2, which provides the breakdown of aggregate income earned by each income class. The proportion remained virtually unchanged during the 1970s. Once into the 1980s, however, the three higher income classes increased their aggregate share while the two lower income classes gradually lost. In 1991, the top 20 percent (the highest income class) accounted for 44.2 percent of aggregate income received, while the upper quarter of this group, the top 5 percent, accounted for as much as 17.1 percent of the total aggregate income.

Table 9.2 Percentage of aggregate income of families (aggregate income by each quintile and top 5% of families)

Year	Lowest (%)	Second (%)	Third (%)	Fourth (%)	Highest (%)	Top 5% (%)
1970	5.5	12.0	17.4	23.5	41.6	15.6
1971	5.5	11.9	17.4	23.7	41.6	15.9
1972	5.4	11.9	17.5	23.9	41.4	15.5
1973	5.4	11.9	17.5	24.0	41.1	15.3
1974	5.4	12.0	17.6	24.1	41.0	15.5
1975	5.4	11.8	17.6	24.1	41.1	15.5
1976	5.4	11.8	17.6	24.1	41.1	15.6
1977	5.2	11.6	17.5	24.2	41.5	15.7
1978	5.2	11.6	17.5	24.1	41.5	15.6
1979	5.3	11.6	17.5	24.1	41.1	15.7
1980	5.1	11.6	17.5	24.3	41.6	15.3
1981	5.0	11.3	17.4	24.4	41.9	15.4
1982	4.7	11.2	17.1	24.3	42.7	16.0
1983	4.7	11.1	17.1	24.4	42.7	15.8
1984	4.7	11.0	17.0	24.4	42.9	16.0
1985	4.6	10.9	16.9	24.2	43.5	16.7
1986	4.6	10.8	16.8	24.0	43.7	17.0
1987	4.6	10.8	16.9	24.1	43.7	16.9
1988	N.A.	N.A.	N.A.	N.A.	N.A.	N.A.
1989	4.6	10.6	16.5	23.7	44.6	17.9
1990	4.6	10.8	16.6	23.8	44.3	17.4
1991	4.5	10.7	16.6	24.1	44.2	17.1
1991–70	−1.0	−1.3	−0.8	+0.6	+2.6	+1.5

Sources: *Statistical Abstract of the United States*, various issues.

Let us next take a brief look at real family income. The median average family income in 1970 stood at $33,238 (in 1990 dollar terms). This figure remained relatively flat until 1980, when it was $33,346, but rose by 6 percent to $35,353 in 1990. On an annualized basis, the average family income grew by a mere 0.5 percent in the 1980s compared to zero in the 1970s. Meanwhile, the highest income class enjoyed the highest rate of growth in income and accounted for a larger proportion of aggregate income.[4]

Finally, Figure 9.3 compares the total net worth of American families in 1983 and 1989. A mere 1 percent of all families accounted for 37 percent of the total net worth in 1989. Moreover, this top 1 percent of families was the only class of income earners that expanded their share of the total after 1983.[5]

Figure 9.3 Share of net worth of American families

Year	Richest 1%	Next richest 9%	Remaining 90%
1983	31	35	33
1989	37	31	32

Percentage of net worth

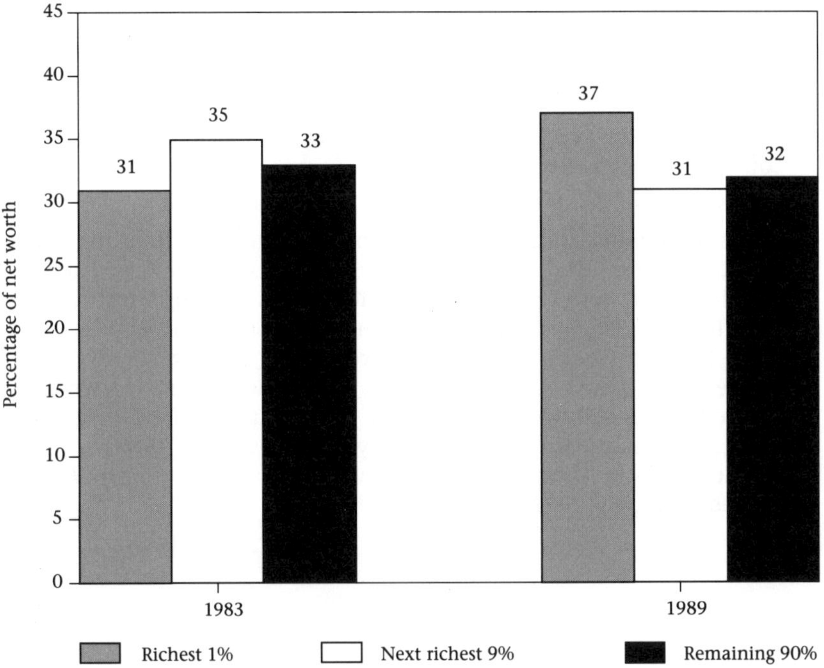

Data: Federal Reserve.

Source: Sylvia Nasar, "Fed gives evidence of 80's gains by richest," *The New York Times*, April 21, 1992.

We now have a clearer picture of the widening income dispersion among American consumers. In Part IV, we will examine how the automakers responded. This will be the last stop in our search for the cause of the U.S. automotive industry's downfall.

Notes

1. *Economic Report of the President,* Washington, DC: U.S. Government Printing Office, 1992, p. 126.

   The Gini ratio is a measure of income differential and is valued between 0 and 1. A Gini ratio near 0 indicates that income distribution is uniform; the closer the ratio approaches 1, the greater the income dispersion.
2. For statistics concerning income dispersion, see Jeannette M. Fitzwilliams, "Size distribution of income in 1963," *Survey of Current Business,* April 1964, pp. 3–11; Daniel B. Radner and John C. Hinrichs, "Size distribution of income in 1964, 1970 and 1971," *Survey of Current Business,* October 1974, pp. 19–31.
3. Money income as reported in this statistic excludes income tax and/or transfer payments.
4. *Statistical Abstract of the United States,* 1992, Table 706.
5. Further examination indicates that some families saw their net worth decline in absolute terms. The largest drops were witnessed among single-parent, non-white, and Hispanic families. The net worth of farm families also declined because of plunging farmland prices. It is ironic that families whose family heads were aged fifty-five or over, and retired, saw their net worth increase, while the median net worth of those families still working decreased. See Arthur Kennickell and Janice Shack-Marquez, "Changes in family finances from 1983 to 1989: Evidence from the survey of consumer finances," *Federal Reserve Bulletin,* January 1992, pp. 1–18.

PART IV

A Vicious Circle

10

Target Marketing

All the clues have now been examined. The time has come for us to present our story.

Figure 10.1 shows three indices: average price of American passenger cars, income levels of the top 5 percent of families, and wage rate for automotive production workers.[1] As you can see, the lines of these three indices almost overlap. (Each index uses 1970 figures as the base rate.) Table 10.1 presents the data on which this figure is based. Of note is that car prices have risen in line with the income growth posted by the top group of income earners.

Figure 10.1 supports the contention that the U.S. automotive industry has indeed targeted only one segment of the market, that is, the highest earning class. The American economy is the most advanced in the world, and the automotive industry seems to have set its eyes on consumers at the very top of that economy.

Let us call this strategy "consumer targeting." It involves developing and marketing products to a selected segment of the market that earns more than a specified level of income.[2] It is worth noting that the industry leadership has long acknowledged that selling to the rich is more profitable. Chrysler chairman Lee Iacocca indicates that automakers have, in fact, shared this strategy. Iacocca wrote that a bigger car can generate ten times the profits of a smaller car, and that the rich are a far more stable class of

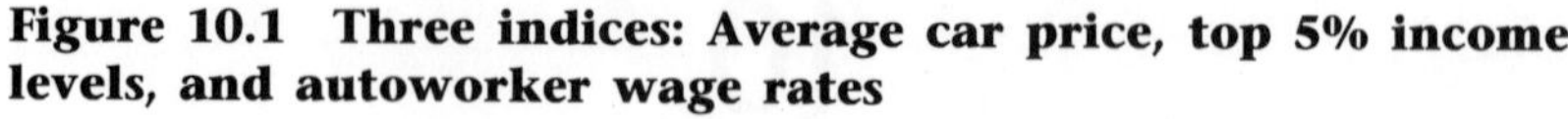

Figure 10.1 Three indices: Average car price, top 5% income levels, and autoworker wage rates

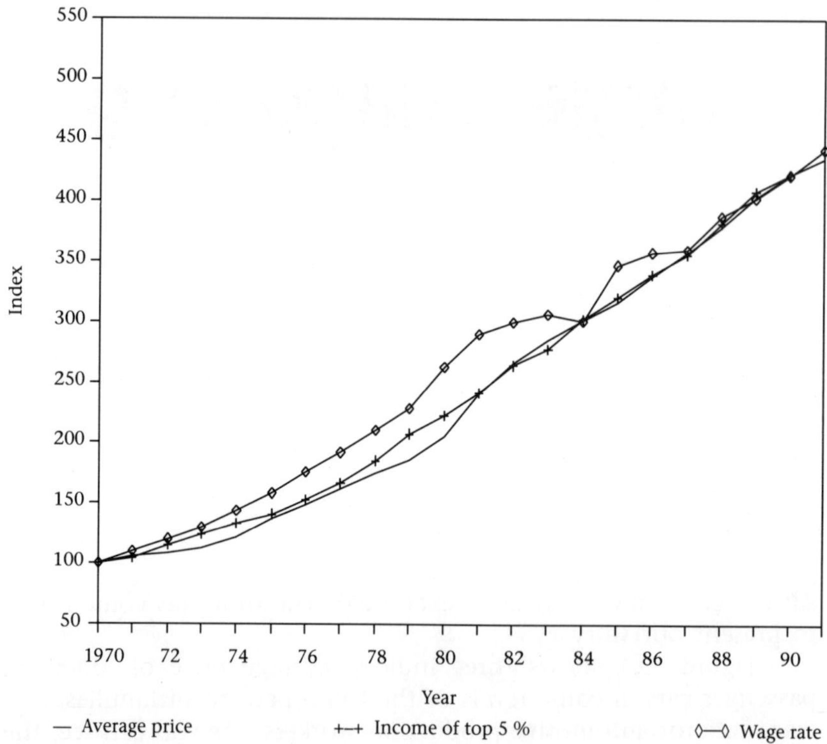

Note: 1970 = 100.
Source: Table 10.1.

purchasers even during an economic downturn because in the U.S. the rich have been getting richer. One of the key success factors in marketing new cars has been to focus on the tastes of the rich.[3]

The rich have thus been seen as a vital market for the automotive industry: the final resort when sales are down. It is possible to tempt the rich into buying new cars with discounts in the form of either price reductions or lower financing rates, regardless of the state of the economy. Moreover, they choose expensive cars loaded with options. The sale of one such luxury car can generate profits equivalent to the sale of ten compact cars.

Consequently, at first glance it appears not a bad idea to target new car development and pricing at this particular segment of the market. This is why automakers have come to base the prices of

Table 10.1 Auto price, top earner income, and industry wage rate

Year	Consumer expenditure per domestic car		Hourly payroll of production workers		Money income of top 5% family	
	$	Index 1970 = 100	$/hour	Index 1970 = 100	$	Index 1970 = 100
1970	3,708	100.0	4.8	100.0	24,250	100.0
1971	3,919	105.7	N.A.	N.A.	25,325	104.4
1972	4,034	108.8	5.8	120.3	27,836	114.8
1973	4,181	112.8	6.2	129.5	30,015	123.8
1974	4,524	122.0	6.9	143.9	32,199	132.8
1975	5,084	137.1	7.6	158.2	34,144	140.8
1976	5,506	148.5	8.5	175.5	37,047	152.8
1977	5,985	161.4	9.2	191.8	40,493	167.0
1978	6,478	174.7	10.1	210.2	44,900	185.2
1979	6,889	185.8	11.0	228.3	50,300	207.4
1980	7,609	205.2	12.6	262.6	54,046	222.9
1981	8,912	240.3	13.9	289.5	58,554	241.5
1982	9,865	266.0	14.4	300.0	64,000	263.9
1983	10,559	284.8	14.8	306.4	67,326	277.6
1984	11,172	301.3	14.5	300.2	73,230	302.0
1985	11,733	316.4	16.7	346.8	77,706	320.4
1986	12,526	337.8	17.2	357.6	82,273	339.3
1987	13,239	357.0	17.3	359.9	86,300	355.9
1988	14,029	378.3	18.7	388.0	N.A.	N.A.
1989	14,957	403.4	19.4	402.9	98,963	408.1
1990	15,641	421.8	20.3	422.9	102,358	422.1
1991	16,150	435.5	21.3	443.8	N.A.	N.A.

Sources: Tables 3.1, 8.1, and 9.1.

various options and models on full-size luxury cars. Since it is only the rich who will buy new cars in a recession, the industry decided to focus on them.

We have already seen that consumers allocate a fixed proportion of their expenditures toward car purchases. This suggests that auto sales should not be affected even if car prices rise in line with income growth among the high-income class. Since this is so, our logic seems to be intact.

To paraphrase the situation, if:

1. consumer income/price of new car = new auto sales, and

2. rate of income growth = rate of price increase of new cars,

then auto sales should remain flat.

This equation is not unconditional. For instance, new car sales may decline if the number of families in this income class falls, even if the income level within this class rises. To boost new car sales in such cases, car prices must be reduced to enable families in the lower-income bracket to afford cars.

Thus far, only a limited amount of research has focused on why income dispersion in the U.S. has widened. All the causes have not yet been identified, but a number of probable reasons may be pointed out. Whether "consumer targeting" increases or decreases sales will depend on the causes of income dispersion.

In this respect, a wider income differential may be regarded as a pertinent phenomenon. Income dispersion—whether the affluent are increasing in number and/or the rich are becoming richer—can be easily observed by automotive market analysts. Real business strategists, however, must look behind the phenomenon and focus on the causes of income dispersion, not merely on the demographical analysis of customers originating in the marketing department. Only in this way can they determine whether targeting the rich is a viable long-term strategy. In other words, marketing strategy must be adjusted over time based on the causes of the income differential change.

How might a change in marketing strategy affect the automotive industry? It is futile, after all, to try to sell the same model to the rich and the not-so-rich. The size of the car, maintenance costs, and the interior of the car, for example, must be adapted to differing lifestyles.

It now takes just under four years to develop a new model. Automakers thus must wait at least four years before they are able to market models adapted to the changes in the marketplace. It is difficult, however, to forecast when and how the factors that induce income differentials may change. Obviously, therefore, it is highly risky to target only one segment of the market when switching from one marketing strategy to another takes so much time.[4]

Before examining the feasibility of income-class targeting as a marketing strategy in the next chapter, let us conclude here by referring once again to Figure 10.1. The relationship among income, price, and wage rates has remained the same since the 1970s—there was no sudden change in the 1980s.

Earlier, after having examined the characteristics of the

automotive industry in the previous chapters, we concluded that some drastic change in the automotive companies must have occurred around 1980. However, another possibility is that the environment surrounding the automotive industry changed. In fact, the basic strategy pursued by the industry has remained unchanged for more than twenty years.

By "basic strategy," I mean the notion of consumer targeting. The assumed objective of this strategy has been to focus on the market segment that offered the highest and the most stable growth. This strategy allowed the automotive industry to post growth in line with economic growth in the 1970s. In the 1980s, however, the environment changed and new car sales began to dwindle. This brought about job cuts and the need to curtail output capacity.

In the 1990s, consumers began to shy away from buying new cars, yet the industry has failed to modify its strategy. The strategy had been effective in the past and become an implicit part of the business.

In the next chapter, we will scrutinize this strategy in more detail. The final chapter in this part will then examine several problems relating to the automotive leadership, problems that delayed their making needed changes in this unsuccessful strategy. The results of these examinations will be used to present some options for the future in Part V.

Notes

1. The automotive sector, as used here, conforms to the definitions used in the SIC code 3711: the sector that manufactures passenger cars, trucks, and buses. These are essentially the Big Three's core businesses. The various categories of auto-related industries, the number of employees in 1991, and their hourly wage rates are as shown in Table 10.2.
2. This argument may be extreme. Note, however, that the weight of high-income earners is rising. Households earning annual incomes of $50,000 or more (in 1987 dollar terms) accounted for 32 percent of new car buyers between 1972 and 1974 but rose to 46 percent in 1990. In the same period, the ratio of households with annual incomes of $30,000 accounted for 25 percent of new car buyers in 1990, down from 32 percent between 1972 and 1974. Households in the mid-income range fell substantially, from 43 percent to 22 percent. These statistics cover both registered owners and principal drivers (for example, students driving cars that their parents have purchased for

Table 10.2 Outline of auto-related industry

SIC code	Industry	No. of all employees (000)	Hourly wage rate of production workers ($/hour)
3465	Automotive stamping	99.3	17.2
3711	Motor vehicles and car bodies	218.1	21.3
3713	Truck and bus bodies	31.1	9.9
3714	Motor vehicle parts and accessories	369.9	15.8
3715	Truck trailers	22.1	10.0
3716	Motor homes	11.8	9.3

Source: U.S. Department of Commerce, *Annual Survey of Manufactures*, 1991.

them). The statistics show that registered owners earn higher incomes than principal drivers.

3. Lee Iacocca, *Iacocca: An Autobiography*, New York: Bantam Books, 1984, p. 90.
4. It is less risky to focus on intangible factors (such as personal taste) than on tangible factors (such as income class). The former strategy is pursued by, for example, Mercedes and Jaguar, both makers of luxury cars. This type of market is smaller than that differentiated by income class, but it is the sort of market that could expand across borders.

11

Rich Man, Poor Man

America radiates an affluent, powerful image: a vast, spacious land of seemingly unlimited natural resources, moderate climate, and enterprising people; a society that emphasizes freedom and creativity, successfully encouraging fresh new ideas. The American economy has grown continuously, supported by such creativity, and the American dream has been inherited down through the generations.

The automobile, electronics, computer, telecommunications, aerospace, and other industries were born in America and have subsequently spread around the world. The new products and technologies emanating from the U.S. have established new industries, led to the formation of giant world-spanning corporations, and generated many job opportunities.

Income dispersion is a natural consequence of such economic activity. Not all families can grow financially at exactly the same pace. Those that have better ideas, or whose members are more industrious or efficient, will reap more rewards for their extra effort. Others will try to emulate such role models, leading to economic growth.

High-income earners tend to be a step ahead of growth. Therefore, an increase in the number of families in this income class implies future economic growth. If a business wishes to expand at a pace faster than the rate of economic growth, it thus makes sense to target this market segment, as the American automotive industry has attempted. The rich, after all, tend to prefer expensive cars loaded with options. The average price of cars will rise if the proportion of such sales rises in relation to total sales. Consequently, overall profitability of passenger car sales will be enhanced. Also, if automakers forsake the rich, these affluent consumers will soon lose their interest in cars and allocate a lesser portion of their income toward car purchases. Thus, there are compelling reasons for the automotive industry to target the rich in order to maintain its position vis-a-vis the overall economy.

More rich people does not necessarily mean fewer poor people. It can suggest an improvement in overall standard of living, and it might merely mean that there is a time lag in income growth in the various income classes. On the other hand, a widening dispersion of income against the background of slower economic growth can imply a deterioration in the standard of living among the less privileged.

The actual GNP indicates that growth in America is slowing down. The U.S. economy grew at an annual rate of 3.8 percent in the 1960s, but this fell to 2.8 percent in the 1970s and to 2.3 percent in the 1980s. The income differential widened in the 1980s, putting pressure on the lowest income class, which saw its real income peak in 1978 and then decline until 1982. Real income growth resumed after that, but it was not until 1988 that the income level returned to that witnessed in 1978.

Meanwhile, the highest income class, which accounted for 41.6 percent of total income in 1980, increased to 44.3 percent in 1990. Wealth has thus become more concentrated in the top 20 percent of all families.

Automakers who had pursued a policy of offering more expensive cars in the 1970s found, in the 1980s, that although the rich became richer, the number of rich families did not increase. The manufacturers could not hope, therefore, to sell expensive cars to more rich families; instead, they had to raise prices in line with income growth. Cars were, thus, becoming more and more expensive. Prices of all cars, from full-size cars to subcompacts, were hiked in this manner.[1]

As far as consumers are concerned, though, a car is just one of many consumable items. It is coincidental that, on the average, consumers have ended up allotting a stable portion of their expenditure toward car purchases. It is natural for consumers to give up their hope of buying a new car if, in their opinion, its price fails to reflect its value.

As we saw earlier (in Chapters 5 and 8), prices of new cars have risen faster than the prices of other consumer goods. A widening difference in the rate of increase was especially evident during the first half of the 1980s; this is the major reason why new auto sales did not expand during the 1980s.

Is it possible for the automotive industry to create buyers of new cars when existing consumers see their income growth level off? Henry Ford attempted to do so in the early days of the automotive industry. In 1914, Ford raised the minimum wage rate of its employees from $2.40 per hour to $5.00 while at the same time reducing the price of the Model T from $850 to $490. Henry Ford expected this strategy to allow his 200,000 employees, and the some 200,000 workers of subcontractors, to be able to afford to buy the car. The Model T was produced en masse with cumulative output amounting to 15 million units.[2]

As Figure 10.1 showed, the wages earned by automotive workers over the past two decades rose in line with the rise in car prices. In 1990, the average employee payroll in the automotive industry (SIC 3711) stood at $42,000, roughly equivalent to the average annual income of families purchasing new cars in 1990. White-collar automotive employees are estimated to have earned an annual income of around $50,000. Automotive companies offer additional benefits amounting to $16,562, on an average, over and above the salaries. Add the two, and the families of automotive white-collar workers are among the top 20 percent of income earners. The annual income of the top 20 percent in 1990 was in excess of $61,490.[3]

New auto sales cannot do other than decline, even when companies pursue this strategy of paying high wages without achieving technological innovation, because demand for other consumer goods rises on a relative basis when the price of cars is hiked. This should lead to reduced job opportunities in the automotive industry and more jobs in other industries. Therefore, the number of consumers who can afford to buy a new car should decrease in spite of the initial goal of automotive companies.

It would not be necessary for us to consider such macroeconomic implications if the automotive industry were smaller. However, since the industry is large enough to exert pressure even on the President, as we saw in Chapter 1, we must consider these types of repercussions.

Numerous white-collar jobs were axed in 1990 and 1991. This reduction was not limited to the automotive industry, but spread to the electronics and aerospace industries. This has reduced the proportion of families in the high-income class, the market segment on which new auto sales depend so much. For those families no longer in the high-income class, the issue becomes one of whether they can afford to maintain their existing cars rather than how much they can spend on a new car. These families were once the major buyers of new cars; they are now required, however, to hang on to their cars for a longer period than they used to. It is natural, therefore, that new car sales have dwindled.

As we have seen, the American automotive industry as a whole has wooed the high-income class. It has aimed to market expensive products to the more affluent and to offer its employees high wages. It has sought to serve the booming segments of the American economy and isolate itself from the other segments. Was such a strategy feasible? Evidence clearly indicates that the answer to this question is "no." While there is no doubt that the rich are important customers for the automotive industry, many problems accompany a strategy that targets only this segment of the market. Let us examine some of those problems here.

Niche Strategy

Targeting a particular segment of the market is basically a niche strategy. Each American automaker sought to capture the same small niche, one too small to accommodate them all. They rushed to seize the same market segment that offered the most stability. The Japanese, who had strenuously developed the low-end of the market in the 1970s to avoid direct competition with the Big Three, were also forced to upgrade and offer higher-priced cars in the 1980s.

It was a situation in which every automaker was trying to become a niche player. They were all competing for the same customers. Competition within a particular category, though, revolves not only around price, but also around design, drivability, and other factors. American automakers were able to improve their capability to compete price-wise, but they lagged behind their foreign rivals in other areas. This led them to lose their market share in the 1980s. The Big Three were cornered as they struggled to compete on the same turf as foreign manufacturers.

Coordination Failure

The reason automakers focused on the high-end of the market was that it offered higher profitability. Although there was a large pool of potential buyers in the low-income class, nobody wanted to serve this market because profitability is low. Also, it becomes even more difficult for low-end manufacturers to emerge when development costs must be increased to satisfy tightened regulatory standards on emission control, fuel economy and safety.

It may be in the interest of a company to concentrate on larger, luxury cars to secure higher profitability. Someone, however, has to take on the task of producing cheaper cars in view of future sales. This is what is referred to as a "coordination failure" in economics, and it is a flaw in the market mechanism.

The analogy of football spectators is often used to explain this. Assume that an exciting game of football is being played. When one spectator in the front seat rises to his feet to get a better view, the people behind him are forced to stand up since their view is now blocked. And then the spectators behind them have to stand up because they cannot see, and so forth. In the end, everybody is standing up, yet they don't have any better view than when they were sitting. The only result is that everyone has wasted energy standing up. (In Chapter 14, we will examine some measures that the government may take to prevent this from happening.)

Domestic Orientation

The use of imported parts and overseas labor may be a way to improve the profitability of cheaper cars. When an automaker decides to focus only on the high-income class, this implies that it is prepared to stake its fortunes on the rich (as the market segment most likely to grow at a faster pace than the growth in the overall economy). All-in-all, consumer targeting is one way to outperform the average growth rate.

Another alternative for growth is for an automotive manufacturer to globalize. Globalization offers great potential for furthering growth in the form of increased exports and/or imports. This benefit comes from being a manufacturer of a full line-up rather than a niche player.

In Chapter 6, we examined past strategies adopted by GM. All of these strategies were lopsided, focusing only on domestic production. The company had no choice but to gamble on technological innovation if it wanted to improve the competitive capabilities of its compact cars while relying only on domestic resources. It is doubtful whether their decision could be justified on economic grounds.

Little Room for Maneuver

It is difficult to change course once a niche strategy is in place. If wage rates are higher than those offered in other industries, it becomes difficult to pursue the lower end of the market. This vicious circle was outlined in Krugman's report (examined in Chapter 3). The only way to maintain the welfare of automotive workers is to maintain high wage rates; yet high wages lead to higher car prices, which in turn puts downward pressure on auto sales. There is no simple means of escaping this vicious circle. The automotive workers themselves must consider the ultimate implications of this cycle. There seems to be no end to job cuts; thus, no one is sure of job security.

An attempt must be made to define the "automotive industry." What is the scope of the industry? Why are automotive employees paid so well? Why is there a wage differential among the various workers within the automotive industry? All of these are questions that must be asked, and answered. Workers can no longer sit back and hope to cling on to their vested interests.

Notes

1. It is difficult to substantiate this assertion since no long-term data is available for comparison. Some sources of data, however, are useful for observing how the annual incomes of passenger car buyers have fluctuated. According to *Surveys of Purchases and Ownership* by the U.S. Bureau of the Census, the annual incomes of new car buyers were in line with the U.S. average until the mid-1970s: the median income of the households who bought new cars was $10,000 to $10,500, while the median income of all households was $10,468 (see MVMA, *Facts & Figures*, 1975, p. 43). By the mid-1980s, however, the median income of households purchasing new cars was more than 43 percent higher than the average; *Newsweek*'s "1986 buyers of new cars" showed the former was $35,600 while the latter was $24,897, and in 1990, this had become $48,095 vs. $29,943, respectively (see MVMA, *Facts & Figures*, 1987 and 1991). It is probable that the shift occurred sometime during this period.
2. Henry Ford, *Today and Tomorrow*, Cambridge, MA: Productivity Press, 1986.
3. Statistics classified by income class are based on monetary income. The data used in the *Annual Survey of Manufactures*, from which annual incomes of automotive workers were taken, is divided into payroll and benefits, which are inclusive of wages, bonuses, and vacation and sick

leave pay. Therefore, their monetary income cannot be accurately isolated from other income. Refer to the Appendix to the *Annual Survey of Manufactures* for a breakdown of other benefits.

12

Management Problems

The Big Three are the world leaders in automotive business management. It was Alfred P. Sloan, Jr., who built GM into the world's largest automotive company. Sloan introduced several marketing methodologies, such as installment sales, annual model change-overs, and fattened sales by loading cars with options. Sloan's contributions enabled GM to overtake Ford, but his greatest contribution was the management legacy that he has left.

Sloan's concept of the organization was unique in the following respects:[1]

- Policy by committee
- Centralized decision-making and decentralized administration
- Balance between financial oversight and operational autonomy
- Guided progression and succession to develop leaders

These organizational principles enabled GM to grow into a giant corporation comprising hundreds of plants, parts manufacturing subcontractors, and dealers, with nearly 900,000 employees worldwide at its peak.

Automotive companies are, in one sense, large-scale manufacturers of a single product. They may produce millions of cars in hundreds of models, but the technology of the manufacturing process is basically the same. All models require procurement, production control, and other functions. This is why automotive companies have the tendency to become enormous. Decentralization is a means of eliminating the obstacles of a giant corporation and restoring corporate vitality; this was the most significant element of Sloan's organizational theory. The delegation of job responsibilities and authority allowed control to be maintained down to the lowest ranks in the organization.

The weakness of decentralization is that employees tend to concentrate on the part of the business that they are directly responsible for; in other words, they lose sight of the overall picture. Financial control and information sharing among committees are mechanisms that can alleviate this problem. Financial control offers the benefit of being a common yardstick against which the performance of each function can be measured. It enables performance to be quantified and compared. This is a systematic manner of managing business, allowing individual departments to make autonomous decisions that are later checked against the financial yard stick.

Special training programs for future management candidates provide them with an overall understanding of the business. Staff with management potential are required to gain experience in various functions so that they will understand how the functions are interconnected. Without this understanding, it would be impossible to envisage the actual business situation from the financial data that the various departments generate. The leadership of giant corporations must be able to see beyond the surface and to understand what the numbers really imply.

Why is a corporation led by such organizational principles unable to creep out of prolonged hardships? If consumer targeting is the problem, why were they not able to identify the problem earlier? This is, after all, a company that has come this far on the rails that Alfred Sloan laid.

This author has no intent to claim that something is wrong with Sloan's philosophy. However, I will point out that the philosophy required of a company that seeks to grow, and that of a giant company seeking to grow further, ought to be different. The issues that GM faces today did not exist in the days of Alfred Sloan.

The first issue is that of size. Suppose that the following pieces of information come up to the leadership at a giant, decentralized organization:

- The marketing department reports that rich consumers are the only ones purchasing cars.

- The employee relationship department reports that the management needs to agree to wage hikes demanded by the workers or risk a major walkout.
- The finance department reports that they will be unable to finance the investment in capital if corporate profits are lower than last year.

Based on this information, what decisions should the leadership take? One answer would be a decision to raise the proportion of luxury cars that the company manufactures. This is the easiest response because it is simply a matter of piecing the information together. But how should the leadership respond if they find that sales do not improve after having pursued this strategy for a while? They ought to question whether the various pieces of information are actually interrelated. Car prices may be raised and wage rates hiked, but we have seen that this "great idea" may actually reduce auto sales and force job cuts. The leadership ought to consider the possibility that the marketing and labor departments are in fact reporting the same phenomenon from a different perspective.

The report from the finance department may be just a reflection of past strategy. It may be a fact that lower profits will make it difficult to finance capital expenditure, but if higher prices contributed toward last year's profits, they should have been able to anticipate slower sales this year. It is their job to review how much capital investment would be required when market conditions change.

This is not an example of losing sight of the overall picture that Sloan cautioned against. Rather, the issue lies in the fact that the company has grown beyond the overall size of the economy. GM has grown so large that its strategy affects the environment in which it operates. Consequently, the individual departments are seeing the repercussions of their own strategies.

The reports that the individual departments file are, therefore, reflections of the actions that other departments have taken, or even those taken in the past by the department filing the report. Merely compiling such pieces of information, therefore, does not give an accurate account of the environment in which the company is operating. The Big Three are "big" corporations and a source of much the nation's wealth. The leadership must be aware of the repercussions of their own strategies on themselves.

The second issue is related to financial control; this is better explained using a specific example. The basic issue is: How valid is the belief that the profitability of full-size luxury cars exceeds that of compact economy cars? It is possible to quantify the profitability of each model by allocating specific overhead costs to each model; this obviously needs to be done to control production costs and to price cars appropriately. But, at the same time, it is natural that profit margins on full-size luxury cars are larger because they are purchased by consumers who earn higher incomes. It is also natural that compact economy cars offer lower profitability because they are sold to lower income earners. But the mission of compacts is to nurture future luxury car users, not necessarily to reap profits. Would it not make more sense to think about overall profits in this manner?

We can think of profits in terms of the whole product line rather than in terms of individual models. The profit on each model may differ, but they all contribute toward overall profits and are all necessary to maintain the business. Cutting off a certain product line because it proves to be less profitable goes against the idea of gaining in the long run.

As we have already seen, the strategy of pursuing the rich has its own risks. There is no guarantee that the income differential will continue to widen over a prolonged period of time, which would favor this strategy. It becomes difficult to consider alternate strategies, however, once a niche strategy is established. Niche players are not free from competition; in some cases, the competition may be even more intense. If automotive manufacturers dwell on the premise that compacts are not very profitable, they will be unable to turn to alternate strategies.

An alternative approach would be to forsake the idea of quantifying profitability by each model. The calculation of profits by model is based on the number of units sold and the retail price, factors that are affected by the economic environment. What sells well depends on specific market circumstances. For instance, compacts sold well after the oil crisis in the first half of the 1980s, while large cars fared well in the latter half (helped by disinflation).

We have also indicated, in Figure 9.1, that the proportion of families in the high-income class fluctuates with the economy. Business will generate profits that are greater than original projections if the overall economic environment is more favorable than anticipated. The manufacturing business is no exception. It is probably more effective to measure the business performance of each model in terms of the market share that it has captured over a competing model, and in terms of the efficiency of production and sales.

It is risky over the long term to shift production totally to models that are profitable at one particular point in time. It would be more effective to have a portfolio of models to diversify the business risk; that is, to manufacture various models with varying profitability levels. It would, however, be necessary to employ a different yardstick, one that is not just financial or quantitative, in order to motivate those involved in the development, production, and marketing of cheaper compacts.

The third issue involves leadership. The strategies pursued by the Big Three reflect, in many ways, the traditional manner in which they have conducted business. They have sought to enhance their competitive capabilities through technological innovation, pursued a strategy of upscaling models and hiking prices, and placed emphasis on the utilization of domestic resources. It is quite clear that their European operations and tie-ups with the Japanese manufacturers have been more successful than their domestic efforts. It is highly likely that the Big Three will play a dominant role in developing the automotive industries in previously communist countries such as China and Russia (a subject that will be dealt with further in Chapter 15 and Chapter 18). For some reason, though, the business has not worked out very well in the U.S.

This is an unsubstantiated view, but the reason may be that alien elements have been ostracized from domestic management. The negative aspects of "guided progress and succession" may have come to rule the organization. On the one hand, shared opinions, viewpoints, and common objectives are important assets for a giant corporation; on the other hand, they leave little room for a "check and balance" system to function properly. The time has clearly come for these corporations to introduce non-traditional thinking into their domestic management.

Note

1. *Call Me Roger*, p. 44. (Reprinted from *Call Me Roger* by Albert Lee, © 1988. Used with permission of Contemporary Books, Inc., Chicago.) See also Alfred P. Sloan, Jr., *My Years with General Motors*, New York: Macfaden Books, 1963.

PART V

Breaking through into the 1990s

13

A Flood of Jalopies: The Hidden Potential

We have so far examined the American automotive industry from two perspectives: that of the manufacturer and that of the consumer. We have seen that the prices of new cars have risen to the point where only a handful of families can afford to buy. While there probably are some ardent car lovers who do not mind spending a major portion of their income on cars, most people need their cars to commute to work or go grocery shopping; for them, prices have risen so much that the purchase of a new car has become a burden. Families that previously replaced their cars every three-to-four years are now doing so every five-to-six years (or even seven-to-eight years). Such a change on the demand side necessarily affects the supply side, that is, the manufacturers.

And it is not only the Big Three that have raised prices; the Japanese and European manufacturers have followed suit. Thus, all domestic and foreign automotive manufacturers have come to concentrate on the same, very limited portion of the automotive market. Further, they compete head-on in all segments of the market, from compacts to luxury cars. Product differentiation (targeting at different users) has become more difficult.

The size of the pool in which the manufacturers swim is becoming smaller. If one swimmer kicks up too much water, other swimmers get splashed, have their passage blocked, or even get kicked. It is no longer possible to swim around the pool freely, oblivious to the other swimmers. We have reiterated again and again that the Big Three are facing hard times, but it is important to note that they are not the only ones suffering. The entire automotive industry is being squeezed into a smaller and smaller pool.

The result? Consumers cannot buy the cars they would like to. The Big Three are scaling back their operations. Workers feel insecure about their jobs. Foreign car manufacturers are finding their options limited. The situation is one in which everyone is suffering from atrophy, and the consequences are proving a substantial burden for American society.

A prime example of the situation is the increased presence of older cars on the roads. Figure 13.1 shows the distribution of cars in use by age. We see from this chart that the number of cars in use that were sixteen years or older accounted for 9.9 percent of all cars in use (12.17 million units) in 1991. This is substantially higher than in the past: 2.9 percent in 1970 and 4.8 percent in 1980.

Figure 13.1 Cars in use by model year

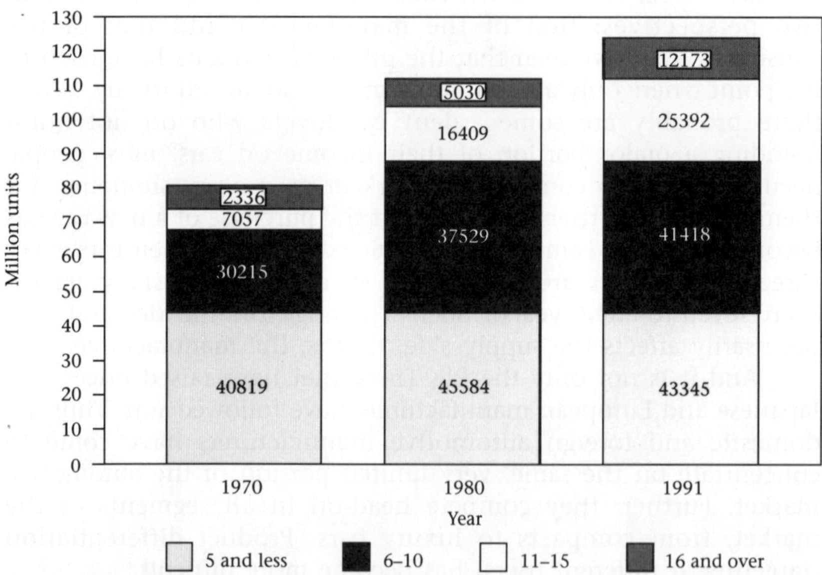

Data: R. L. Polk & Co.
Sources: MVMA, *Facts & Figures*, various issues.

Let us assume, for the moment, that the percentage of these older cars in 1991 had been 4.8 percent (the value for 1980), or 5.92 million cars. Subtracting this from the actual number in use (12.17 million) yields a difference of 6.25 million cars that could have been replaced with new cars. In other words, auto sales in the 1980s should have grown by 625 thousand units annually. We saw in Chapter 2, for example, that the average number of cars sold per year was 10.2 million in the 1970s and 9.8 million in the 1980s. If older cars had been replaced with new cars at the same rate as in the 1970s, annual auto sales would have been higher in the 1980s than in the 1970s.

If consumers did not have to drive over-aged cars, the Big Three would not have had to curtail their operating capacity, and auto workers need not have worried about losing their jobs. The consequences of the increased presence of these old cars goes far beyond this, however. The continued use of old cars has great impact on many issues that exist today.

One such issue is fuel consumption. It is well-known that the U.S. federal government regulates the fuel economy of new automobiles to reduce the country's dependence on imported oil and reduce oil exploration to preserve the environment. The current regulatory standard is 27.5 miles per gallon (mpg). The actual numbers achieved for 1992 models are 26.9 mpg for domestic cars and 29.0 mpg for imported cars. The average comes to 27.8 mpg, which is higher than the federal standard. However, this is only the average for new cars. The actual average, when used cars are taken into account, is much lower, as shown in Figure 13.2.

The average distance traveled per vehicle in 1990 (inclusive of both new and used cars) was 10,556 miles. A total of 72,435 million gallons of gasoline were consumed during the year.[1] This indicates that the average fuel economy was only 20.9 mpg, or just 76 percent of the federal standard. Fuel economy improved by 38 percent in the 1980s, but, even at this rate, it will take until the year 2001 to achieve the federal standard of 27.5 mpg.

In 1990, 50.2 percent of all oil used in the U.S. was consumed by automobiles running on the highway. This amounts to 20.7 percent of all energy consumed in the U.S.[2] Therefore, improving the fuel economy of used cars will contribute a great deal toward conserving energy. Tightening regulatory control on new cars alone is not enough if over-aged cars are not replaced with new cars.[3] If auto sales remain at current levels, it will take some fifteen years to achieve the current federal standard.

The same may be said of emission control and safety regulations. Emission control was stepped up in 1990 with the

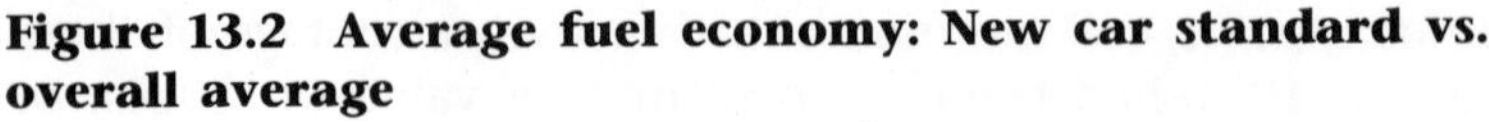

Figure 13.2 Average fuel economy: New car standard vs. overall average

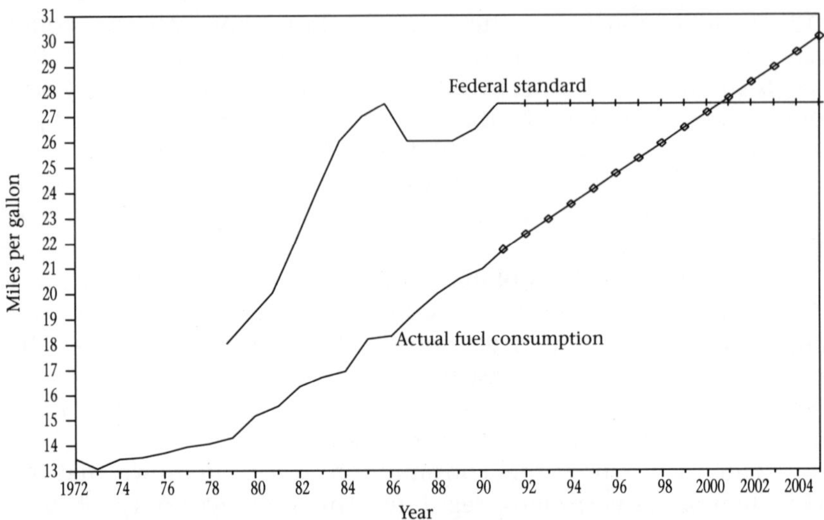

Note: Estimates by K. Ohmura.

Data: U.S. Department of Transportation, Federal Highway Administration.

Source: MVMA, *Facts & Figures*, various issues.

enactment of the Clean Air Act, which aims to cut toxic emissions by 50 to 85 percent of the current level. As a first step, manufacturers are required to reduce the output of nitrogen oxides to 60 percent of the current level and the output of hydrocarbons to 40 percent by the time the 1994 models are introduced to the marketplace. The emission control equipment is required to last for at least ten years or 100,000 miles. In some states, such as California, where air pollution is a serious problem, the law requires the gradual introduction of automobiles powered by motors other than gasoline engines.[4]

In the area of safety, too, old cars cannot comply with current regulations. The National Highway Traffic Safety Administration decided in 1990 to introduce new test methods to ensure that cars are sufficiently reinforced to withstand side crashes. The installation of side-restraint air bags is also being considered.

There is no doubt that steps in each of these regulatory areas will help the automobile to become a safer and more environmentally friendly machine. However, it is not sufficient to regulate new cars without regulating the over-aged cars if the regulations are to be effective within a short period of time.

The basic need is to find a means of promoting the sale of new cars—in part, to satisfy the social need to improve the environment, increase safety, and reduce energy consumption. The suppliers, which include both the Big Three and foreign car makers, need to come to grips with these issues.

The vast presence of used cars implies the existence of tremendous potential demand for new cars. Think of it in these terms: more than 12 million cars are ready to be scrapped! The strategies employed to capture this potential demand will determine the fate of the automakers: that is, who will survive. It is clear that the American automotive market can still be revitalized; it all depends upon how the manufacturers undertake this enormous challenge.

Notes

1. U.S. Department of Transportation, Federal Highway Administration, "Highway Statistics"; quoted in MVMA, *Facts & Figures*, various issues.
2. U.S. Department of Energy and U.S. Department of Transportation.
3. This issue was first raised around 1980. See Robert W. Crandall et al., *Regulating the Automobile*, Washington, DC: Brookings Institute, 1986.
4. "The Greening of Detroit," *Business Week*, April 8, 1992; and *Ward's Automotive Yearbook*, 1991, p. 30.

14

Advanced Standard Strategy

Before examining how American automakers might go about replacing over-aged cars with new cars, let us take a look at the role of the government. The business climate could deteriorate even further if the government decides to step up regulatory controls on environmental protection and in other areas. In other words, tighter controls increase development costs, which pushes up the price of new cars, which in turn puts downward pressure on auto sales.

It is estimated that the cost of satisfying environment-related regulatory standards is between $2,000 and $4,000 per car (1994 model). This accounts for some 12 percent to 25 percent of the average price of domestic cars ($16,150 in 1991). The U.S. Bureau of Labor Statistics estimates that it cost an additional $2,717 per car (in 1991 dollar terms) to install the required safety and emission control features during the twenty-five-year period from 1968 through 1992. It will cost the manufacturers even more to meet future regulatory standards. If these extra costs are passed through to the consumer, as they must be, auto prices will be hiked at the same rate that they have risen between 1968 and 1991. It is necessary to promote the replacement of over-aged cars with new cars when tightening regulatory controls on new cars in order to improve the current environmental situation.

Normally, the government might resort to the following measures under these circumstances:

- Provide tax incentives for automobiles that meet regulatory standards.
- Offer tax deductions and cheap financing for investment in facilities that produce the devices required to meet the regulatory standards.
- Hike the cost of owning over-aged cars to encourage users to replace them with new cars.
- Provide incentives to scrap over-aged cars.
- Raise the tax on gasoline.
- Offer incentives to promote the development and/or use of cleaner sources of energy.

In other words, the government would normally take measures to reduce the price of new cars and/or raise the cost of over-aged cars. Alternatively, it could reduce the cost of owning a new car so that it becomes cheaper than owning an old car.

However, since the American government suffers from a massive budget deficit, it cannot take measures that would result in a financial burden. Also, since the wealthier households tend to own new cars while the poorer households own old cars, raising the cost of owning over-aged cars might force lower-income families simply to give up their cars rather than purchase new ones. The government, thus, faces a dilemma and has been unable to take specific measures in this area.[1]

The most efficient way of dealing with the problem would be to take gradual steps, building upon a series of well-planned measures. First of all, the government should make clear that it expects to meet regulatory standards in the least costly fashion, then take measures to streamline the cost of development. It should look to promote joint research among automakers in certain technological areas to avoid possible overlap in investment. In other areas, it should monitor the cost of using patents so that the manufacturer who develops a new technology does not overcharge others for its use.

In some areas, the government could promote competition by rewarding automakers that are able to develop technologies for meeting or exceeding regulatory standards while penalizing those that are unable to do so. The current method of applying uniform standards obligates the manufacturers to develop whatever technology is necessary on their own. This functions merely to raise costs, which ultimately get passed on to the consumers.

A second area that needs to be tackled is the reduction of manufacturing costs. The use of imported parts should be allowed, especially in cases where locally procured components push up manufacturing costs. The notion of "imports" in this context should be understood in the broadest sense of the word: they will include the use of Mexican labor, Japanese manufacturing know-how and/or capital, Asian parts, and any other resources that may help to lower the cost of production. It is important to note that the increased sales that will result from lower automobile prices will lead to increased job opportunities, not only among auto-related manufacturers, but among dealers, finance companies, and other businesses.

It is clear that increased production will lead to more job opportunities. The use of foreign resources will not conflict with the need to enhance America's competitive capability in the international marketplace. Look at Japan, which relies heavily on imported natural resources (including oil and minerals) but has secured for itself a strong position in the international market. The U.S. should learn from its experience in the 1980s; protecting domestic manufacturers does not, over the long-run, help to secure jobs.

Third, the U.S. could strengthen its competitive position by making its regulatory standards the world standard. To achieve this, the regulatory standards must be such that they are able to gain worldwide approval. Whether the American regulatory standards on fuel economy, environmental protection, and safety will burden the automotive industry must be examined from a global perspective. Europe and Japan are moving toward establishing similar standards, and interest in controlling air pollution is mounting in Central and South America as well as in Asia. The U.S. should endeavor to maintain its leadership by refining the rationale behind its standards so that they are more persuasive.

Finally, price formulation must be monitored. Recent findings by the Economic Strategy Institute (ESI) indicate that the Big Three's cost of manufacturing a compact can differ by as much as $1,526 (19.3 percent) per car, depending on whether they are operating at full capacity or at 80 percent of capacity (which was the case in 1991).[2] Obviously, costs decline as production increases.

If governmental measures are taken in the future to stimulate auto sales—such as encouraging the replacement of over-aged cars by offering cheaper new cars, which will lead to increased output —the benefits, that is, the lower manufacturing costs arising from enhanced capacity utilization, should not be monopolized by the automakers. A part of the gain should be plowed back to the consumers.

The government should also monitor dealers to ensure that they are offering the cheaper cars to consumers. The current tendency among dealers is to load cars with options to raise the sales, and to sell to the customers with higher purchasing power. There are cases in which consumers are denied the opportunity to purchase cars at the (cheaper) prices shown on published price lists. It is important that dealers be made aware of the social benefits that accompany the sales of less expensive cars in large volumes. Dealer margins should not be set in a manner that is disadvantageous for promoting the sale of compacts.

The most important factor in all this is that the government's stance be consistent. High regulatory standards need to be met at the lowest possible cost in order to benefit society as a whole. For this, existing measures should be combined and implemented in a flexible manner so that the ultimate objectives are achieved.

The goal of regulatory standards will not be achieved quickly if the government merely sets the numbers and leaves the method of meeting them up to the automotive industry. The lessons of the 1980s have taught us that the government must take the initiative and must participate fully in the implementation of regulatory standards.

Notes

1. The introduction of the Junker Car Proposal, recommended by the National Automotive Dealers Association, seems to have been considered at one point. Owners of older cars would have been paid $700 to $1,000 for voluntarily turning in their old car. Companies that bought the old cars could avoid penalties for other forms of air pollution that they produce. The idea was proposed to help low-income earners to replace their over-aged cars with new ones. However, this contradicted the original intention (of cleaning up the air) as it provided the buyer with a loophole. See "Bush gives auto industry break on vapor recovery," *Automotive News*, March 16, 1992, p. 36.
2. Clyde Prestowitz, Paul Willen, Larry Chimerine, Kevin Kearns, and Robert Cohen, with Mike Flynn and Sean McAlinden, *The Future of the American Auto Industry: It Can Compete. Can It Survive?* Washington, DC: Economic Strategic Institute, 1992.

   This report compares the manufacturing costs of major companies in the U.S. and Japan. As of 1992, assuming full-capacity operation and an exchange rate of 130 yen per dollar, Ford's manufacturing cost is the lowest, followed by Chrysler and Toyota in that order. The average manufacturing cost of the Big Three is lower than that of the four major Japanese automakers (Toyota, Nissan, Honda, and Mazda), primarily because the costs of purchased components and materials are lower for the American car companies. Labor productivity (man-hours per car) is more efficient in Japan by 34 percent, and wage rates are 19 percent lower.

The report, however, assumes that the costs of purchased components and materials are the same for the four Japanese automakers; this is hard to believe. Also, it is not very clear on its treatment of research and development costs. Overall, though, it is highly likely that, in dollar terms, the difference in the manufacturing costs between American and Japanese automakers has narrowed since the first half of the 1980s. As we have seen, Japanese makers can no longer compete solely through cheaper prices.

Productivity and labor costs are important factors in determining future cost differentials. Japanese automakers plan to reduce the average number of work hours by 10 percent from 2,265 hours a year in 1989 to less than 1,900 hours by 1995. Automation through the introduction of industrial robots cannot be achieved overnight. Its introduction has resulted in a plateau of productivity in terms of the number of vehicles produced per employee. Meanwhile, wage rates are constantly rising. The same tendency may be observed for component suppliers.

In the U.S., meanwhile, capacity utilization is rising and manufacturing costs have been falling. This trend is expected to continue through 1993–94. Therefore, it is unlikely that the Japanese manufacturers' cost-competitiveness will improve in the near future.

Foreign exchange fluctuations greatly affect manufacturing costs in the short-term. However, over the long-term, foreign exchange rates reflect not only the general price levels in a country, but also labor productivity. Therefore, it is important that the American automotive industry raise its productivity at a faster pace than the industrial average to become comparatively more competitive. The impact of exchange rate fluctuations will be reduced if wage rates can be determined by productivity gains (Hiroshi Yoshikawa, "On the equilibrium yen-dollar rate," *American Economic Review*, June 1990, pp. 576–83).

15

Direction of the Big Three

Our examinations in previous chapters have indicated that if the Big Three continue to pursue existing strategies, they will find that auto sales (and hence production) will continue to dwindle. These existing strategies call for raising car prices at the same rate as the growth in income among the rich and increasing industry wage rates in line with the increase in car prices. What other options are available to the industry?

The difficulties of altering strategy are many. Low-income earners comprise the group most affected by the fluctuations in the economy, and low-end cars offer low profitability. If wage hikes are constrained while products are upgraded, it will lead to discontent among the union members. Labor strikes can create substantial losses, including reduced market share. (GM suffered from a series of walk-outs during the first half of the 1970s, which enabled Ford to catch up.)

For many years the strategy of consumer targeting worked; manufacturers did not need to look for alternatives. But the economic environment changed in the 1980s. If the same strategy were pursued today, auto sales would fall, and jobs would be cut. The final blow to strike the industry was the recession that hit in 1990–91. The time has clearly come for automakers to adopt a new strategy.

With these points in mind, let us examine some of the options that are open to the automotive industry. The matrix in Table 15.1 is useful for examining consumer targeting in relation to other strategic options. The first axis indicates whether a company focuses on a specific market segment or on all segments. In terms of the product line-up, these two options can be expressed as either "full line-up" or "niche." The second axis indicates whether the manufacturer will depend solely on domestic resources or whether it will take advantage of foreign resources. Put more simply, it tells us whether the automaker has decided to be a totally "domestic" company or an "international" one. Although simple, these classifications help us to identify the range of strategies that the automotive manufacturers could take.

Table 15.1 A strategic matrix of automotive manufacturers

Resource/Market	Product portfolio	
	Full line	**Niche**
International	Import/Joint venture	Export
Domestic	Protection	Specialized market

At one time, all three U.S. companies offered a full line-up of products built using only domestic resources. Their cars came in all sizes and prices. This began to change in the 1970s, though, when car prices were raised and the manufacturers began to focus on the wealthier segment of the market. Recent statistics indicate that cars costing $10,000 or under account for less than 10 percent of total sales. Thus, though automakers seemingly continue to offer a full line-up, in fact they have shifted toward a niche strategy of pursuing consumers with higher incomes. In terms of the aforementioned matrix, the Big Three would be classified as "domestic" and located somewhere between "full line-up" and "niche."

It is up to each manufacturer to decide whether it wishes to continue to offer a full line-up or pursue a niche market. All three U.S. makers need not pursue the same strategy. Pursuing the niche market is just one option; they are not being forced to choose it over the other. In view of the current economic outlook, it may turn out to be more advantageous to offer a full line-up.

First, it is difficult to envisage the income differential widening further. The issue is gaining mounting interest and was taken up in the presidential campaign. American society could be split further if the income differential is allowed to increase or even remain as it is.

Second, there is less competition worldwide in the low-end of the market. Japanese automakers entered the U.S. market on the strength of their lower costs in the 1970s, followed by the Koreans in the 1980s. The Big Three were not competitive in this area since they were not used to manufacturing small cars. Recently, though, the Japanese and Koreans have begun to lose their competitive edge, for such reasons as the stronger yen and rising labor costs.

Manufacturers in developing countries, on the other hand, lack the technology to meet the stringent regulatory standards of the U.S. Cheaper prices are impacted if expensive high-tech components must be imported from developed countries. The Big Three will enjoy the benefits of raised operating capacity once consumers begin to replace their over-aged cars with new ones, and they may once again find themselves to be the world's most efficient automotive manufacturers.

Whether a company decides to be "domestic" or "international" is a matter of calculation. The decision will be based on which choice will help to reduce costs and whether there is an overseas market for the products that the company produces. Until now, the Big Three did not have the freedom to make this choice because of social pressures to "buy American." Corporate relationship with the UAW was also a source of difficulty. It is clear that the leadership of the three companies preferred to use domestic resources.

However, a decision to manufacture all models using domestic labor and parts, which are more expensive than the international average, will become a serious constraint in the future. Taking advantage of cheaper overseas resources to manufacture cheaper cars need not reduce jobs in the U.S. over the long-run. If manufacturers are willing to nurture the low-end of the market, it will eventually develop buyers of more expensive cars in the future.

The North American Free Trade Agreement (NAFTA) represents an important change in the economic environment, one that will affect the automotive industry. Manufacturing in Mexico could lower costs substantially. The typical wage of a worker who belongs to the automotive union in that country was $3.12 per hour in 1989;[1] a UAW worker earns as much as $21.51, almost seven times the Mexican rate.

According to a report by the ESI, fifty to seventy man-hours are required to build one car. This suggests that the cost per car would be reduced by between $920 and $1,379 by producing them in Mexico, a reduction of between 14 percent and 22 percent in the price. The Big Three have an edge over both the Europeans and the Japanese when it comes to entering the Mexican market; taking advantage of the situation could open up new possibilities for the Big Three in the 1990s.

Probably the most effective strategy for the Big Three will be to take advantage of foreign resources and manufacture low-end cars, thereby expanding demand. In terms of the matrix in Table 15.1, they should choose to offer a "full line-up" and seek to become "international." The sticker price on passenger cars that would be bought as replacements for over-aged cars will need to be in the range of $6,000 to $6,500.[2]

The cheapest models offered in 1992 were the captive imports, which sold for between $6,579 and $6,999. This means that the prices of low-end cars will need to be reduced by about $1,000 from current levels. These low-end models would have simple interiors and very few options; there will be little room, price-wise, for fancy electronic gadgets. One concept for such a model will be discussed in Chapter 19.

The efficient use of advanced technology and cheap labor are prerequisites to developing such low-end cars. The competitive capabilities of the Big Three should be enhanced, since they will be producing devices to meet emission, fuel economy, and safety standards in large quantities. They also will be able to use cheaper Mexican labor to manufacture some of this equipment. The Big Three have been exposed to Japanese manufacturing know-how through NUMMI and other joint ventures with the Japanese; they are thus well equipped to develop the low-end of the market.

It should be noted at the same time, however, that there are certain risks associated with developing cheap cars. It will not be feasible to change models frequently as in the case of luxury cars; low-end cars will need to be marketed in limited versions and over a longer period of time. But if a rival firm begins offering frequent model changes, this will be difficult to ignore. The use of foreign resources also implies increased currency and national risks. Moreover, the Big Three are not accustomed to the concept of pursuing the low-end of the market; they may lack sufficient development know-how in this area.

Tie-ups with Japanese or European automakers would be an efficient means of overcoming these weaknesses. There are already some examples of this: Ford's Hermosillo plant in Mexico, for example, is based on Mazda's Hofu plant and has the highest quality control of all Ford plants.[3] It would not be out of place for GM or Chrysler to consider teaming up with a foreign manufacturer.

It needs to be confirmed here that the introduction of cheap cars is tantamount to adding a new category of cars. The rich who have been buying luxury cars will continue to buy those cars; these cheaper models will sell to consumers who could not previously afford to buy a new car. The idea is to encourage them to replace the over-aged, polluting cars that they are driving. In terms of profitability, however, luxury cars will continue to provide the larger source of profit. The low-end cars will contribute little in comparison to larger cars, though they will potentially lead to increased sales of luxury cars over the long-run.

This being the case, it is important to consider how to win market share in this category at the lowest possible cost and risk. The Big Three could, obviously, try independently to take on this task; they need not open up new possibilities for the Japanese. Teaming up with the Japanese may offer benefits, however, in light of the risks associated with investing in an area that will offer limited short-term rewards.

The reason that the Big Three should look to dominate the low-end of the market is not limited to expanding their share of the American market; such a strategy could also have a significant impact on developing new demand worldwide. The demand for automobiles rises sharply once per capita GNP/GDP exceeds $2,000 to $3,000; prime examples of this are Brazil, Mexico, Taiwan, and South Korea. The age of motorization also will soon dawn in Eastern Europe and Russia, if these countries are able to make the shift to a market economy. The retail price of Russian-made cars is set between $2,000 and $5,000. The average price of used cars in the U.S. is $7,550. Russia and Eastern Europe currently import used cars from the West, but if new cars are priced low enough, they could replace these used cars.

China, too, will be entering the era of motorization early in the twenty-first century. The Big Three could hold the trump card in developing this market. Both Russia and China are similar to the U.S. in that they cover a vast land area. Both may find larger cars more comfortable. These cars will also need to travel over long distances and be adapted to the cold climate.

The former communist countries will require help in developing their automotive industries; they will need partners to team up with. The U.S. is in a more advantageous position than either the Europeans or the Japanese in this regard. The organizational set-ups and systems of American companies are more comprehensible to the outsider than either European or Japanese ones. It is highly likely, therefore, that we will see a division of labor, so-to-speak, in this area: a combination of American and European automakers, or American and Japanese automakers, offering technology and capital, with local markets supplying local labor and demand.

It is possible that the American automotive manufacturers could come to play a pivotal role in the world marketplace, since they already encompass a potentially enormous domestic market for cheap cars. Manufacturing costs will be reduced sharply through mass production if the export market is added. The Big Three ought to pursue the strategy of offering a full line-up and of becoming more international to revitalize themselves on a global scale.

The other strategies indicated on the matrix offer only partial or temporary remedies. "Domestic" and "full line-up," for example, implies that the automaker will continue to supply the market with a full line-up of cars under the current price structure. However, compacts and low-end cars offer low profitability, and there is no guarantee that an automaker will be able to maintain its competitive position in the global marketplace over the long-run.

Protective measures, such as restraining imports or restricting the transfer of plants to Mexico, would be required under a domestic, full line-up strategy. It would be difficult to lower the current level of prices without substantial compromise on the part of the unions. Under such a strategy, the manufacturer would be supplying a full product line in terms of size, from small to large cars, but the main consumers that it will be targeting would be the wealthier segment of the market. It is a strategy that seeks to maintain the status quo; ultimately, it will lead to reduced sales.

The "domestic" and "niche" strategy is probably the least risky. However, it does not offer a market big enough to sustain all three U.S. automotive manufacturers. The most stable automotive market is not the market for passenger cars; it is the market for the specialized vehicles used by the military, rescue and ambulance services, and fire departments. In the private sector, there is less foreign competition in general cargo vans and food service vans. The same is true of government cars, such as police patrol cars.

Fleet cars for rental and lease are unprofitable markets because of fierce competition among the Big Three and foreign car penetration. The luxury car market depends heavily on income growth of the high-income class, and it is doubtful whether this market segment will continue to remain the most stable passenger car market in the U.S.

The aforementioned markets are limited in terms of the number of cars that can be produced and sold. However, the Big Three could position themselves in this area, since they rely heavily on the profits reaped from this market segment. Future domestic production and, consequently, the source of profits for the Big Three can be expected to shift from passenger cars to trucks and other specialized vehicles. This option could supplement the passenger car business.

"Niche" and "international" is the strategy that European automakers like Mercedes and Volkswagen are pursuing: internationalization of the luxury car market. To achieve this, a manufacturer must be equipped with sufficient merchandising capabilities to provide for the consumer with discerning tastes. However, competition in this area in Europe and Japan is already quite fierce. The Big Three would be required to make a huge investment in new model development and in establishing sales channels. The Middle East is a potential market, but the overall number of countries with strong demand for luxury cars is limited. Therefore, it will be difficult for the Big Three to take on this strategy overall, although individual product lines such as Cadillac and Lincoln could potentially proceed in this direction.

In short, the most practical strategy would be for the Big Three to concentrate on redeveloping the low-end of the market in the 1990s, building on the lessons that they have learned during the previous two decades. It is this strategy that takes advantage of the fact that America is the largest market in the world, with the largest economic potential. If this strategy goes well, the Big Three may find themselves in a position to once again lead the world automotive industry in the coming century.

Notes

1. Kristin Fitzpatrick, "Booming Mexico industry awaits decision on free trade," *Ward's Automotive Yearbook 1991*, pp. 112–13; "Detroit south," *Business Week*, March 16, 1992, pp. 16–21.
2. Based on the following calculations: The average expenditure on domestic cars in 1990 was $15,682, or 15 percent of the average income earned by the top 5 percent (that is, $102,358). Fifteen percent of the median income earned by all families in the same year ($35,353) is $5,303. If cars are priced at this level, then average families will be able to purchase new cars. In the middle of the 1990s, this is estimated to increase to between $6,000 and $6,500 in 1990 dollar terms.
3. "The partners," *Business Week*, February 10, 1992, pp. 102–7.

16

Japanese Manufacturers' Options

It is probable that the relative profitability of the American market will decline for the Japanese automotive manufacturers. Already, cars sold in the U.S. are cheaper than those sold in either Europe or Japan. Japanese cars may be expensive commodities for American consumers, but their prices are comparatively cheaper (in yen terms) because the value of the dollar has fallen. This means that the American market is no longer a very attractive one for the Japanese automakers.

Also, in dollar terms, market growth is not very high. Any attempt to capture a further share of the market will be very costly. If the Japanese are unable to profitably export more cars to the U.S., then they must increase the output of their transplants to gain a greater market share. They will also have to establish parts-manufacturing plants alongside their assembly plants because of tighter local content regulations. All-in-all, the Japanese automakers would have to make new investments in the U.S. while their factories back home operate at reduced capacity.

One Japanese manufacturer, Daihatsu, has already announced its intention to pull out of the U.S. market. It would not be surprising to see others follow suit. Even the larger manufacturers may decide to review their investments in the U.S. and instead begin to shift their focus to Asia. The Japanese automakers' interest in penetrating the American market has already reached a saturation point.

Those companies that are able to expand the operating capacity of their transplants will be the ones that survive the competition among Japanese manufacturers. We should not expect to see all Japanese makers scrambling to expand their plants in the U.S., however. Some are already suffering from low-capacity utilization. Even if some of the automakers decide to increase their operating capacity, they may decide against building additional plants. Rather, they may look for optimum solutions, among which might be:

- acquiring plants that have already been built by other Japanese automakers,
- acquiring or jointly using currently unused parts of the plants owned by the Big Three, or
- importing vehicles from plants in Central and South America, Europe, or Asia.

There has been mounting concern that the expansion of transplants may become the next major trade issue between the U.S. and Japan, but it is unlikely that such expansion will take place.

The major issue on which the Japanese automakers will focus is surviving the competition with other Japanese manufacturers in the global market. The Japanese car makers compete in the same market segments. It is, therefore, difficult for them to develop a niche. Thus, their primary mission is one of maintaining the market share that they have captured; to achieve this, they must ensure that their businesses remain profitable.

One lesson learned in the 1980s was that consumers in different parts of the world have different tastes and preferences. Automobiles suited to the American market differ from those suited to the European or the Japanese markets; this is why GM's project to build a "world car" dwindled into obscurity. The Honda Accord, for example, became one of the most popular cars in the U.S., but it does not sell well in Japan. Mercedes are used as taxis in some countries because of their solid construction, but they are considered luxury cars in the U.S. and Japan.

What these examples tell us is that automakers must develop cars adapted to the needs and preferences of each local market. Earnings from the local markets will finance future research and development. The manufacturers that can develop products adapted to the local marketplaces will survive in the international market. Those that cannot do so will be unable to escape the confines of a single market.

There are nine passenger car manufacturers in Japan. Of these, few are competitive internationally, with Toyota, Nissan, and Honda being the strongest candidates. Mazda may be regarded as a part of the Ford group. Mitsubishi lacks a firm foothold in the U.S. and Europe; its major playing ground is Japan and Asia. Suzuki, Daihatsu, and Fuji Heavy Industries supply mini-cars and have the potential to penetrate the Asian and East European markets, but they are likely to seek closer ties with GM, Toyota, and Nissan, respectively. GM-affiliated Isuzu is giving up its passenger car business to specialize in trucks. The current structure of the Japanese automotive industry is thus approaching its end and is in the process of being consolidated.

From a global perspective, Toyota is gradually strengthening its relationship with GM and Volkswagen. To keep up with this move, Nissan will probably look to strengthen its relationship with Ford and Mazda. We can expect to see the relationship between Honda (which is very strong in the U.S. but rather weak elsewhere) and Mitsubishi (which is weak in the West but strong in other markets) stepped up in the future.

For Japanese automakers, the American market is their second largest (after the Japanese market). But as far as the 1990s are concerned, having a large share of this market may not necessarily be an advantage. On the contrary, a large presence in the U.S., in terms of operating capacity, could become a heavy burden for manufacturers.

Let us look at a typical example of the problems that can be encountered. In the January–March quarter of 1991, Honda decided to reduce output at its Ohio plant for the first time in its history. This decision was made in light of the need to adjust inventory as sales plunged when fighting began in the Gulf. Honda decided that production would be cut by 3 percent from the initial plan.

However, the result of this move to reduce production slightly was that Honda's manufacturing costs soared by a much larger margin than originally anticipated, and employee morale declined. Employees had formerly been given various incentives to increase productivity on the floor: awards like free trips for those who had submitted good proposals for *kaizen* (improvement) and those who had increased productivity. Such opportunities were lost when output was reduced, adversely affecting employee incentive; the output of Honda's most productive employees fell by the largest margin.

The attitudes of Honda's parts and material suppliers also changed. Suppliers who had done their very best to maintain top-notch quality, with a view to securing more orders in the future, began to get careless. Believing they saw a red light blinking in the plant's future, they allowed deliveries that contained faulty lots.

The Japanese system of manufacturing is effective only when all those involved try their very best. It is a delicate system that does not function well if someone decides not to chip in. Honda's experience shows that there can be unpleasant side-effects if everyone's job is cut back by the same proportion, although this is considered to be the fairest way in Japan. They have learned that, in the U.S., it may be more effective to lay off some employees and have the rest continue to work at the same pace.

Well-managed plants need to be able to adjust or cut output when required. It is impossible to be profitable over the long-run if plants cannot change the models or the number of vehicles that they manufacture based on fluctuations in auto sales and changes in consumer preference. The challenge facing the transplants is to improve their profitability rather than to expand their capacity. Future foci will be on increasing exports to Europe and Asia, producing a wider variety of models, and raising the local content ratio.

Finally, a word on how the Japanese automakers now view the low-end of the market. They believe that high prices have caused auto sales to fall in the U.S., and they have drawn up a number of plans to introduce new cars at prices that will encourage consumers to replace their over-aged cars. These plans have not been put into practice, however, because Japan is bound by an export quota; manufacturers have preferred to export upscale cars since those are more profitable.

The idea of manufacturing cheap cars at the Japanese transplants was also considered but was determined not to be financially feasible. One concern was that their increased presence would re-ignite trade friction between the two countries.

The strong yen has emerged as an additional factor that hinders the introduction of cheap cars. Heavy labor costs are also thought to be an obstruction. The dollar is expected to trade around the 120 yen range or higher for the next few years, so the Japanese automakers are considering the possibility of manufacturing low-end cars overseas (that is, outside of Japan). Several Asian countries, Mexico, Spain, and Eastern Europe are possible candidates for such production, with Mexico as the strongest candidate for supplying the North American market.

It is likely, in that event, that the Japanese automakers will consider forming joint ventures with American car makers or even collaborating among themselves, rather than risk braving it alone. Like the Big Three, the Japanese manufacturers see the need to diversify their risk. Future increased market share will be won by those who can introduce low-end cars to meet the demand in the market.

It is difficult to respond immediately if the question is raised as to what are the particular strengths of Japanese cars. The Japanese manufacturing systems and quality control are now the only ready answers; previously, it would have been possible to say that those strengths were represented in a car like the Toyota Corolla. Today, though, we see cars such as the Accord, the Miata, the Camry, and the 300ZX—cars that offer different strengths.

Probably the greatest strength of the Japanese car makers lies in their ability to offer features suited to each market segment. Do the Japanese have an edge over their European or American counterparts? It is true that the Japanese automakers have contributed a great deal toward modernizing the manufacturing process, but the world's other automakers are fast catching up. The Japanese have reached a stage where they must consider whether and how they can keep a step ahead of the global automotive industry, and what they can do to achieve this goal.

PART VI

Lessons from the U.S. Automotive Industry

17

Status of the World Automotive Industry

I once again visited the market specialist of a Japanese automotive company in early March 1993. My main purpose in this interview was to clarify the picture of U.S. automotive demand. In 1992, retail sales of passenger cars in the U.S. rose slightly (by 0.5 percent) from the level of 1991, to 8.21 million units. Sales of trucks increased by 12.4 percent, to 4.91 million units. In the fourth quarter of 1992, especially, consumer confidence exhibited a remarkable improvement.

A second point of interest was the market share of Japanese vehicles in the U.S. market. The market share of Japanese passenger cars decreased from 30 percent in 1992 to 27 percent in February 1993. It was alleged that Japan's international competitiveness was becoming weaker because of the yen's appreciation.

My final topic of interest was the world automotive demand and supply situation. A view of the prospects for 1993 suggested that the profitability of the Big Three would recover, while the earnings power of Japanese and European manufacturers was likely to deteriorate. There were serious discussions among automotive analysts as to whether a recovery in the U.S. demand might support global demand.

I started the interview by looking back on the developments in the U.S. automotive market in 1992. "Well, do you think the market has already hit bottom?"

"In the case of trucks, yes."

"How about passenger cars? There has been a surge, hasn't there, with retail sales in December up 11.3 percent from the previous year?"

"The year-over-year increase is good," the market specialist replied, "but the seasonally adjusted annualized rate is still poor. In the fourth quarter it was 8.07 million units, which is virtually flat from 8.03 million units of the third quarter. As you know, this is quite a low level. And it fell to only 7.8 million in February. We don't yet see any significant increase in the passenger car market."

"A very low level, indeed," I agreed.

"Yes, the most recent peak was 8.36 million units in the second quarter of 1992. The market has been slowly decreasing since then."

Looking at documents on the table between us that tabulated marketing data analysis of the retail sales of cars and trucks in detail, I observed, "But truck sales are good, aren't they?"

"Yes, truck sales are good; very good. The seasonally adjusted rate for the fourth quarter of 1992 was 5.28 million units, and the trend has been steadily rising from 4.5 million units in the first quarter."

"In that case, how would you describe the overall trend in the U.S. market?"

"The market is recovering, but quite weakly," the analyst responded. "Basically, the slow growth of the overall economy is affecting the automotive market. Looking at the real GDP growth rate, for example, the fourth quarter of 1992 showed a 2.9 percent increase from the previous year. In the fourth quarter of 1983, however, the corresponding period for the last recovery, the rate was a much higher 6.7 percent. And the unemployment rate is still at 7.3 percent. Overall, I'd have to say there has been no significant improvement in consumer spending for non-service consumables."

The documents on the table also contained a comparison between the 1983 recovery and 1992, from a macro-economic point of view. These documents showed such factors as real GDP growth rate, the unemployment ratio, consumer spending, and personal income. I questioned the specialist about the demographic analysis of the passenger car market.

"Can you give me any further details regarding the car purchasers?"

"Of course. The most remarkable feature in 1992 was that personal demand increased while corporate demand decreased. Personal demand rose 4 percent year-over-year, but corporate demand decreased by 10 percent. A change in the manufacturers' policy regarding program cars is an important factor in this. Without the effect of the decreased demand for program cars, the total demand for cars in 1992 would have been larger by some 150,000 units."

"Second, among individual buyers, the most active classes of purchasers by occupation were clerical/sales employees and blue-collar workers. These groups increased their purchase of cars by 19 percent and 24 percent, respectively, from the previous year. On the other hand, professionals and retirees stayed away from the market. Professionals aged twenty-nine and younger and retired people sixty years of age and older bought 12 percent fewer cars in 1992."

"As a reflection of this trend," the specialist continued, "in the West, such as California, where the unemployment ratio was at a high 9.2 percent, sales were virtually stagnant. In the Southwest and South, however, sales were relatively strong."

"So, after all," I concluded, "no significant change has occurred in the market since our last meeting, has it?"

I felt at ease as I listened to the expert's story, because it corroborated my own hypothesis that the U.S. automotive demand was undergoing a structural change. Improvement in the unemployment rate was the key factor that would stimulate an increase in demand. In order to decrease unemployment, production would have to go up throughout the economy. Consequently, automotive demand would follow the recovery of output in other sectors. This was a reverse of the phenomenon of the past, when the motor vehicle industry was the nation's leading industry.

"That's right," the analyst agreed. "We also feel that our analysis has been confirmed. Obviously, the replacement cycle has become much longer. The average changeover period was 5.0 years in 1989; it was 5.7 years last year. Based on this figure, we estimate that the number of cars in use probably fell in 1992."

"Is it true? The number of cars in use is falling in the U.S.?"

"It hit a peak in 1991. I think that 1992 should be the year when the number of cars in use started to decline."

"Countries in which the number of cars in use has dropped are quite rare in the world," I noted.

"Quite rare indeed," the expert confirmed. "But, it is significant that the number of trucks in use has been rising; we think this probably will cancel out the fall in the number of cars. Overall, then, the total number of vehicles in use is not decreasing."

According to R. L. Polk & Co.,[1] the number of cars in use on July 1, 1991, was 123.32 million units. This is virtually the same as the previous year, 123.27 million units. On the other hand, the number of trucks in use increased from 56.02 million units to 58.18 million units during the same period. If the number of cars in use should show a decrease in 1992, however, this would mean that the diffusion rate of passenger cars in the U.S. fell for the first time in history.[2] One might conclude from this that some class of U.S. families had become unable to keep even used cars. The distortion of income dispersion, discussed in Part IV, thus seems to have affected the economy much more severely than ever.

I changed the topic to the demand for trucks. "Why do you think the demand for trucks is so strong?"

"Well, among light-duty trucks, the most active segment is the sport-utility vehicle. Chrysler's Cherokee and Ford's Explorer are in this segment; many purchasers of vehicles in this segment probably have shifted from passenger cars. These are high-income professionals with annual incomes of over $75,000. The styling and design, which is differentiated from passenger cars, are appealing to such customers."

"The main buyers of pickup trucks, on the other hand, are blue-collar workers with annual incomes of between $30,000 and $50,000. They have shifted from passenger cars to pickup trucks because of cheaper prices. And probably the major van purchasers are family-oriented people. The most representative occupation of this segment is clerical workers, with an average income of between $50,000 and $75,000. These van buyers think it is important to travel together with family members."

"So, after all," the analyst concluded, "the customers who purchase light trucks are spreading over a wide range. These purchasers prefer light trucks to cars not only for economic reasons, but also based on their lifestyles and personal tastes. For them, light trucks do not belong to a different category from cars. Rather, I think that our definitions of cars and light trucks have become outdated."

As Figure 17.1 shows, the average shipment prices of trucks has been rising more slowly than car prices since 1980. This is the economic factor behind the strong sales of trucks as compared with cars. Aside from this, though, consumers also seem to be attracted by the different characteristics. As already mentioned in Part II, truck sales have been increasing since 1980, so the demand shift from cars to trucks did not come as a surprise to the marketing expert.

Figure 17.1 Average shipment price index: Passenger cars vs. trucks

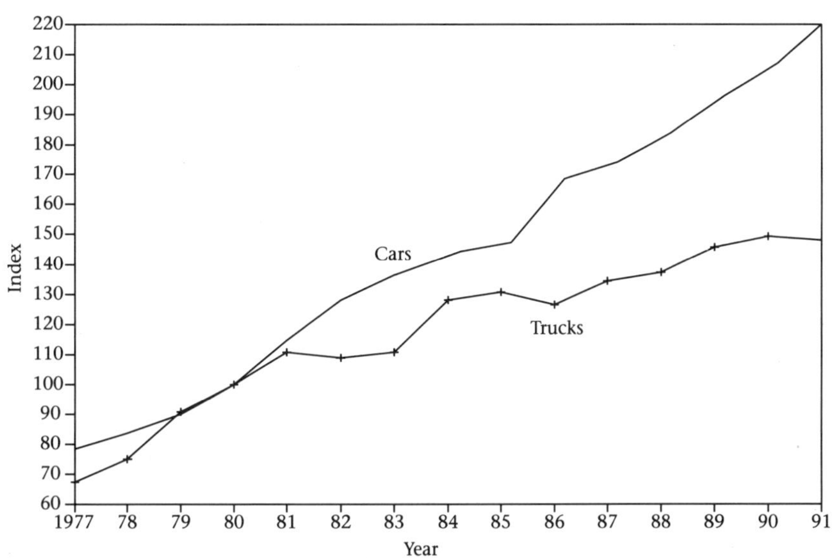

Note: 1980 = 100.

Sources: Department of Commerce, Bureau of the Census, *Annual Survey of Manufactures*, various issues; Department of Commerce, Bureau of the Census, *Census of Manufactures*, 1982 and 1987.

I next turned the topic to market share issues. "Do you think," I asked, "that total sales of vehicles are a better reflection of the current sales momentum than the car sales trend? That showed a 5 percent growth in 1992."

"That's right. At any rate," the analyst cautioned, "the 1992 recovery was much slower than that of 1983. Actually, in 1983 the retail sales of trucks jumped by 22 percent year-over-year, and the total sales of vehicles increased by 16.8 percent."

"The truck market is dominated by domestic manufacturers, isn't it?"

"Yes. Domestic truck sales grew by 16 percent in 1992, while Japanese truck sales actually dropped by 6 percent. As a result, the market share of the Japanese manufacturers decreased from 17 percent to 14 percent."

"Was the same trend evident in the passenger car market?" I inquired.

"Well it's apparent that Japanese competitiveness in the U.S. has been weakening in relation to domestic cars, especially vis-a-vis Ford and Chrysler. But the Japanese market share in 1992 slipped only by 0.1 percentage point."

"Did price hikes hamper the Japanese manufacturers?"

"Perhaps. The average price increase in domestic cars in 1992 was 1.4 percent, compared with a 5.4 percent surge in prices of Japanese cars. On top of this, domestic manufacturers continued to offer more sales incentive programs, while Japanese manufacturers decreased their incentives."

"A price comparison of the Toyota Corolla with the Chevrolet Corsica may make matters clear. In 1991, the Corolla was priced at $9,418. This rose to $11,198 for the 1992 model, making it almost as expensive as the Chevrolet Corsica. However, the Corsica belongs to the class that is one rank higher than the Corolla. For example, the length and width of the Corsica are 4,658 mm by 1,732 mm; in comparison, the Corolla is just 4,369 mm by 1,684 mm. Normally, we would expect the Corsica to be about 15 percent more expensive than the Corolla."

"So the price competitiveness of the Japanese manufacturers is now quite weak."

"Yes, but it is not only an issue of pricing. In terms of quality, too, the Japanese manufacturers are facing tough competition. For example, the Saturn is now ranked third in consumer satisfaction research. A recent study also shows that Chrysler's LH gets a higher evaluation for styling than does Toyota's Camry, which is the newest model among Japanese cars.[3] It is not an easy job to develop cars that can overcome these quality and design considerations; this is the consensus among our engineers, I believe."

"So the Japanese cars have lost their competitiveness even in terms of design and quality?"

"I'm not sure whether our competitiveness is reversed, but we can definitely say that the Japanese no longer lead the market in any sense."

I had a strong interest in the pricing policies of the Japanese manufacturers, as the yen-to-dollar rate had continued to appreciate since the beginning of 1993. If Japanese prices were to be raised without the domestic manufacturers following suit, the Japanese market share would inevitably drop even further.

"Then, you cannot raise the retail prices any more, can you?"

"Well, we have already increased the selling price by 1.2 percent, on the average, this February. Ford raised their prices by only 0.4 percent, and Chrysler by just 0.2 percent. So we are seriously concerned about our market share. In February 1993, the Japanese market share dropped to 27 percent, compared with 30 percent last year. It is possible that this will fall to somewhere around 25 percent by the year end."

"Now, the yen-dollar exchange rate is about 115 yen per dollar. Do you think the current export prices reflect the foreign exchange of this level?" This was a crucial question, because the Big Three were reportedly about to file dumping charges against Japanese manufacturers in January. Unless Japanese retail prices were raised sufficiently to account for the appreciating exchange rate, the difference between Japanese domestic prices and U.S. retail prices would definitely expand, providing the Big Three with concrete grounds for taking legal action in the future.

"No, I don't think so. We have already hedged the foreign exchange rate by buying futures at 120 yen per dollar for the first half of this year. However if the rate remains higher than 115 yen per dollar in the second half, our prices ought to be raised. Models lacking non-price competitiveness will be chased out of the market."

Apparently, outdated Japanese models have already been defeated in the marketplace. For example, in January and February 1993, Honda's Accord sold only 57 percent as many units as in January and February of 1992; three years have already passed since the model was first marketed. The honor of best-selling car was recaptured in 1992 by the Ford Taurus. Even in the Japanese-dominated segments like subcompacts and compacts, the Saturn now threatens Corolla, Civic, and Sentra. Once Ford and Chrysler market small cars that are as competitive as the Saturn, the Japanese status as most popular cars will be in jeopardy. It is reported that Saturn will reach the break-even point sometime in 1993.[4]

"So the Big Three already hold the initiative in the market-place?"

"Except for GM. Because GM is not yet as strong as the other two, the Japanese market share can be maintained at this level. With GM's improvement, however, it will undoubtedly be lowered even further. We believe that now the U.S. automotive situation converges on GM's condition."

"GM seems to be overly dependent on larger cars. Isn't that its major problem?"

"Certainly GM has suffered from the stagnant demand by retired people. But it's even more seriously damaged by a weak cost-competitiveness. GM lags far behind Ford and Chrysler in this regard."

An article that reports the ranking of domestic automotive plants based on labor productivity bears out this contention.[5] In the case of passenger car plants, the top four places are monopolized by Ford, with a Ford and a Chrysler plant tied for fifth and a Chrysler plant in seventh place. The fifteen GM plants are ranked between thirteenth and twenty-ninth place. In truck plants, also, Ford dominates the top seven places, followed by three GM plants. Chrysler's six truck plants are less competitive than its passenger car plants, ranking between thirteenth and twenty-third place.

GM's market share decreased to 34.6 percent in 1992, down from 35.6 percent in 1991. In the second half of the year, it dropped further, to only 31.7 percent. In January and February, 1993, it was 26.4 percent and 26.8 percent, respectively.

GM chairman Stempel announced his resignation at the end of October 1992. Mr. Smale became the chairman and Mr. Smith was promoted to CEO in early November. The new management then announced a further downsizing plan, one that called for the closure and sale of ten plants, changes in the organization of the Cadillac division, and the relocation of employees in the affiliated subsidiaries such as EDS and GMAC. However, Moody's Investment Services downgraded GM's bond rating to Baa1 from A2.[6] GM's quick revival appears unlikely.

In summary, the current U.S. market appears as follows. First, overall demand is recovering very slowly. Especially, passenger car demand has not yet shown any significant improvement. Second, among new car buyers, purchases by white-collar workers, such as professionals within the high-income class, and retired persons are remarkably stagnant. Third, Japanese cars are losing their market share because of weak price competitiveness and the catch-up in quality by U.S. domestic cars. Finally, among the Big Three manufacturers, GM's market share is plummeting.

"Considering the current situation, what is your prognosis of demand in 1993?"

"We estimate that passenger cars will sell slightly less than 9 million units; trucks will reach 5.0 to 5.1 million units. In total, then, the demand will be 14 million units. This is not a pessimistic forecast, but rather an optimistic one; taking into account the effect of the introduction of a gasoline tax and/or an increase in the income tax for high- and middle-income classes, the forecasts probably should be revised down. We are not sure about President Clinton's economic policy yet. In the medium-term, however, we believe that the passenger car demand will never surpass 10 million units. Even including truck sales, total vehicle sales can scarcely reach 15 million units in the next peak year. It is likely that the record high of this century has already been achieved in 1986."

Although the Big Three manufacturers have to some extent regained their competitiveness vis-a-vis Japanese manufacturers, they will be unable to expand their production capacity under such an uncertain demand situation. If they upgrade their product mix, or raise the prices of their cars in the face of a capacity shortage as they did in 1985 and 1986, this will obviously cause a heavy demand fall in the next economic downturn. Furthermore, easy concessions in wage rate negotiations in a prosperous phase would weaken their cost-competitiveness once the foreign exchange rate grows stronger in terms of the U.S. dollar versus the Japanese yen.

These two points leave it open to question whether the Big Three have really experienced a turn-around or not. Profits, no matter how big they might grow, are not in the long-term a good index of industry robustness. After all, the Big Three will have to take radical steps to deviate from the past locus. The same thing is now true, however, of other manufacturers in Europe and Japan as well.

The specialist finally mumbled, "Nevertheless, the U.S. market might be rather good in 1993 in comparison with other markets. The European and Japanese markets are likely to be much more miserable."

The subject then changed to Japanese and European markets. Passenger car demand in Japan dropped by 8.5 percent, and that of trucks and buses also fell by 5.7 percent in 1992. This is the first time that the Japanese market has experienced a decrease over two consecutive years. Decelerated growth in personal income, mainly due to the cut in overtime work, has adversely affected consumer confidence.

In the medium-term, there is another factor that will cause the growth rate of cars in use to slow down: due to a declining birth rate, the number of Japanese reaching age eighteen (the minimum age for getting a car or truck driver's license) will consistently decrease, from the current 2.1 million to only 1.2 million in the year 2010. The growth rate in the number of valid driver's licenses was around 5 percent annually in the 1970s and about 4 percent in the first half of the 1980s. This fell to 3 percent or so in the second half of the 1980s and is expected to approach 2 percent by the turn of the century.

If we compare the diffusion rate of vehicles per licensed drivers, the Japanese rate in 1990 was 947.3 vehicles per 1,000 licensed drivers (versus 1,134.9 vehicles per 1,000 licensed drivers in U.S.). However, in 1970, the U.S. rate was 973.2 vehicles, much nearer to the Japanese rate in 1990. Japanese demand for vehicles will suffer from cyclical ups and downs depending on the economic environment, just as the U.S. demand has done since 1970.

In 1992, the Japanese automotive industry exported 45 percent of its products. Of the total 5,668 thousand exported vehicles, 32 percent were sent to the U.S., 21 percent to the EC, and 47 percent to the rest of the world. However, passenger car exports to the U.S. were under a voluntary restraint of below 1.65 million units in both 1992 and 1993. Also, exports to the EC are restricted to under 1,089,000 units in 1993 (down from 1,185,000 units in 1992).[7] Since the historical high of exports in 1986 at 6,730 thousand units, they have decreased for seven consecutive years, falling to 5,668 thousand units in 1992.

The production of vehicles in Japan in and after 1993 hinges on the increase in the domestic demand and exports to the rest of the world, mainly Southeast Asia and the Middle East. However, only a 1–2 percent growth in the domestic demand is expected, and the decrease in exports to the U.S. and the EC is unlikely to be offset by the increase in exports to the rest of the world. This is the basic reason why the Japanese market specialist I spoke to is so pessimistic about the demand and supply situation of the Japanese industry.

European demand for vehicles is also seen decreasing after 1992. Although last-minute buying of cars in Germany and the Netherlands supported 1992 sales in the EC at the same level as 1991, the EC Committee expects 1993 demand to decrease by 6.7 percent. In January and February 1993, retail sales of vehicles in the eighteen major European countries plummeted 12.9 percent and 17.9 percent from the previous year, respectively.

Table 17.1 tabulates the global trend of vehicle demand. One remarkable finding is that 1980 and 1991 were abnormal years in the sense that all three advanced economies (the U.S., EC, and Japan) experienced a simultaneous fall in demand. In years other than those two, the decrease in some countries was almost canceled out by the increase in other countries. A second fact to notice is the high growth rate in the developing countries. Between 1980 and 1992, developing countries enjoyed 63.8 percent growth, while U.S. growth rose only 14.5 percent, Japan 38.8 percent, and the EC 48.6 percent. Among EC countries, Portugal, Greece, Spain, and Italy lead the expansion of the automotive market. The increase in their sales of vehicles during the same period was 251.2 percent, 138.3 percent, 80.6 percent, and 52.8 percent, respectively.

Table 17.1 Global vehicle demand

Year	World	(YOY)	U.S.	(YOY)	EC	(YOY)	Japan	(YOY)	The rest
1980	33,825	−9.3	11,466	−18.9	9,488	−5.0	5,016	−2.7	7,855
1981	33,049	−2.3	10,796	−5.8	9,449	−0.4	5,126	2.2	7,677
1982	33,763	2.2	10,542	−2.4	9,679	2.4	5,261	2.6	8,280
1983	35,360	4.7	12,312	16.8	9,797	1.2	5,382	2.3	7,868
1984	38,536	9.0	14,484	17.6	9,622	−1.8	5,437	1.0	8,993
1985	41,771	8.4	15,724	8.6	10,784	12.1	5,557	2.2	9,706
1986	43,087	3.2	16,322	3.8	11,844	9.8	5,708	2.7	9,214
1987	43,664	1.3	15,189	−6.9	12,744	7.6	6,018	5.4	9,712
1988	46,993	7.6	15,679	3.2	13,407	5.2	6,721	11.7	11,187
1989	47,858	1.8	14,713	−6.2	13,943	4.0	7,257	8.0	11,945
1990*	48,036	0.4	14,146	−3.9	14,136	1.4	7,777	7.2	11,977
1991	46,269	−3.7	12,539	−11.4	14,084	−0.4	7,525	−3.2	12,121
1992	47,153	1.9	13,124	4.7	14,201	0.8	6,959	−7.5	12,869
1992/80	139.4		114.5		149.7		138.8		163.8

*After 1990, EC includes the demand in the former East Germany.

YOY = Year-over-year change.

Sources: Nippon Jidousya Kougyoukai (Japan Automobile Manufacturers Association, Inc.), *Syuyoukoku Jidousya Toukei* (*Automotive Statistics for Major Countries*), various issues; Table 2.3.

The United Kingdom, Germany (excluding the former East Germany), and France show 0.5 percent, 31.0 percent and 12.3 percent growth, respectively. Of the developing countries in Asia, Korean sales jumped from 104 thousand units in 1980 to 1,268 thousand in 1992, and Taiwanese sales jumped from 131 thousand units to 541 thousand units.

In sum, the growth of global automotive demand has come to depend much on the developing countries. Moreover, in the future, the market in the advanced countries is unlikely to grow as fast as in the past ten years; Japanese demand seems to have become already saturated, and demand in the U.S. will be on a downward trend. Thus, it is the developing markets that should be focused upon.

Is there any possibility that manufacturers in developing countries will emulate the Japanese manufacturers of the past by making use of the motorization of their home country? Such companies should have not only cost-competitiveness but also sufficient quality in terms of fuel efficiency, emission control, and safety. They will need to clear high standards of regulations to penetrate into the advanced markets.

In the 1980s, no such manufacturers appeared in the global market. Korean manufacturers like Hyundai were supposed to follow the Japanese, but they failed to maintain their position in the U.S. For example, the import sales of Korean cars in the U.S. dropped from 264,282 units in 1988 to 102,029 units in 1992. Technological competitiveness, especially in electronics components and metallurgy, will continue to be a high hurdle for the advanced countries' rivals to overcome.

The problems of the U.S. manufacturers are now spreading to the European and Japanese manufacturers. European and Japanese manufacturers can no longer rely on the advanced markets. However, neither can they enter into the developing markets because of the lack of low price capability and government import restrictions in some countries.

This dilemma is similar to the problem U.S. manufacturers face in the U.S. in many respects, especially in that there is great demand for new vehicles, but the advanced manufacturers cannot cope with such needs. In the following chapters, possible solutions will be discussed, drawing upon the lessons gleaned from the analysis of the U.S. automotive market.

Notes

1. "Yearly passenger car scrappage and growth in the U.S.," *Ward's Automotive Reports*, March 30, 1992, insert.
2. Actually, cars in use on June 30, 1992, fell 2.98 million units from the previous year to 120.35 million units, while trucks in use increased 2.99 million units in the same period to 61.17 million units. The total number of vehicles in use showed a slight growth. See "Yearly passenger car scrappage and growth in the U.S." and "Yearly truck scrappage and growth in the U.S.," *Wards Automotive Reports*, August 16, 1993, insert p. 2.
3. Richard Homan, "Popularity contest," *Road & Track*, December 1992, pp. 79–96.
4. Chrysler has recently reported that it will market a new subcompact car from early 1994, Neon, which will be quite competitive with Japanese small cars both in price and quality. Neon's prices will range from $8,600 to $12,500, which are lower than Honda's Civic and Nissan's Sentra, although the car will include driver's and passenger's air bags as standard equipment. See "Chrysler's Neon," *Business Week*, May 3, 1993; "Saturn enters 'break-even year'," *Wards Automotive Reports*, January 11, 1993.

5. John McElroy, "Ranking the assembly plants," *Automotive Industries*, January 1993.
6. According to Moody's rating, "bonds which are rated A possess many favorable investment attributes and are to be considered as upper-medium-grade obligations. Factors giving security to principal and interest are considered adequate, but elements may be present which suggest a susceptibility to impairment some time in the future." On the other hand, "bonds which are rated Baa are considered as medium-grade obligations (that is, they are neither highly protected nor poorly secured). Interest payments and principal security appear adequate for the present, but certain protective elements may be lacking or may be characteristically unreliable over any great length of time. Such bonds lack outstanding investment characteristics and in fact have speculative characteristics as well." In these ratings, "the modifier 1 indicates that the security ranks in the higher end of its generic rating category; the modifier 2 indicates a mid-range ranking; and the modifier 3 indicates that the issue ranks in the lower-end of its generic rating category" (*Moody's Global Rating*, April 1993, p. 216).
7. The EC Committee and the Japanese government agreed to reduce exports to the EC further to below 1 million units in the autumn, in view of the continuing decline in automotive demand.

18

Dynamics of the World Automotive Industry

Let us now consider a generalized model of automotive industries around the world, a model that is derived from our previous considerations of the U.S. automotive industry. This model concentrates on the relationships among the average price of vehicles, average production cost of manufacturers, and affordability by the consumers. Through use of this model, we can better understand the implications the U.S. industry presents for other manufacturers.

We will start by examining the relationship between personal income and the diffusion rate of vehicles. In Part II, we saw that American households have spent a relatively constant portion of their income on automotive-related items. This implies that, as income rises, the demand (measured in terms of total number of vehicles in use) will increase. The central issue here is whether or not this relationship can be extended to other countries in a different stage of development.

Figure 18.1 shows the conclusions of a cross-sectional analysis for 1987. The horizontal axis indicates the per capita GNP or GDP of 127 countries, plotted on a natural logarithmic scale. The vertical axis measures the number of vehicles in use per thousand of population, also on a natural logarithmic scale.

Figure 18.1 Per capita GNP/GDP and vehicle diffusion rate

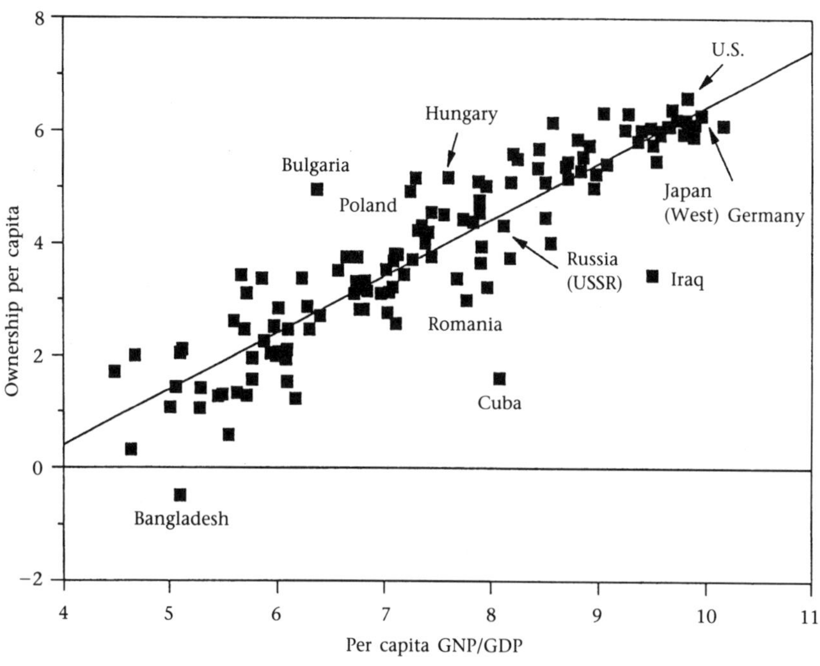

Per capita GNP/GDP = log (per capita GNP/GDP [$])

Vehicle diffusion rate = log (number of vehicles in use / population [units per 1,000 people])

Vehicle diffusion rate = $-3.625 + 1.006 \times \log$ (per capita GNP/GDP)

Determinant Coefficient = 0.817

Data: Japan Motor Manufacturers Association, *Syuyoukoku Jidousya Toukei (Automotive Statistics of Major Countries)*, 1989; International Monetary Fund, *International Financial Statistics*, 1991.

In the most advanced countries, such as the U.S., the former West Germany, and Japan, national per capita income is high, ranging from more than $18,000 to $26,000. These countries also have a high diffusion rate of vehicles, ranging from 408 vehicles per 1,000 population for Japan and 492 per 1,000 for the former West Germany to 734 per 1,000 for the U.S.

In the middle of the figure are nations such as the former Soviet Union ($3,364 per capita income and a vehicle diffusion rate of 76 units per 1,000 population), Taiwan ($4,983 and 88 per

1,000), Korea ($2,729 and 39 per 1,000), Brazil ($2,312 and 85 per 1,000), and Mexico ($1,724 and 96 per 1,000). The average for all the 127 countries is $3,471 per capita income and 107 vehicles per 1,000 population.

Those nations at the left side of the distribution are the very poor countries, such as Bangladesh ($163 per capita income and less than 1 vehicle per 1,000 population) and Ethiopia ($104 and 1 per 1,000). Because both axes are measured by a logarithmic scale, Bangladesh's diffusion rate is shown as a minus value in Figure 18.1.

An ordinary regression shows that these two factors correlate very well. The determinant coefficient is 81.7 percent (as indicated in the notes to Figure 18.1). Based on this data, the income elasticity of demand throughout the world is estimated to be around 1.006. This means that a 1.000 percent increase in per capita income would cause demand for vehicles to grow by 1.006 percent. If the replacement cycle of vehicles is constant over time, therefore, the demand for new vehicles will grow at the same rate as the growth of the number of vehicles in use. We can thus conclude that, in the long run, the demand for new vehicles has a positive relation with the national per capita income.

It should be noted that the above proposition applies to "normal" countries; that is, countries in which the use of vehicles is not limited by political and social constraints. There are several exceptions to this, however. Although countries like Bulgaria, Poland, and Hungary show a rather strong popular preference for vehicles, the data might be biased to some extent by these nations' intent to promote the relative advancedness of their economies vis-a-vis their Western counterparts. These countries are famous for the use of over-aged cars and trucks. Then there are Muslim countries such as Iraq and Iran that limit driver's licenses to males for religious reasons. And there are some cases, such as Cuba, in which per capita income is probably overestimated.

Now that we have established the relationship between per capita income and the number of vehicles in use, let us consider the relationship between per capita income and the average price of new vehicles. Looking back to the replacement cycle of vehicles discussed in Parts IV and V, we saw that new vehicle prices are a prime determinant of the length of this cycle over the long run. If the prices of new vehicles are too high (in comparison with consumers' incomes), the use of over-aged vehicles will prevail and the cycle will become longer. Alternatively, if prices of new cars are relatively low, consumers will scrap their over-aged vehicles to purchase new ones.

Thus, it is apparent that the average new vehicle price is related to per capita income, as shown in Figure 18.2. The line labeled "affordability" shows this relationship. An affordable price is the price level at which consumers can afford to buy new vehicles within a relatively short period to replace the old ones. Even though people want to achieve a better environment by replacing over-aged cars with new ones, the cost of it should be within the limit of the average income. Therefore, the affordability line has a positive coefficient in relation to the per capita income.

Figure 18.2 Economic development and the automotive industry: Normal case

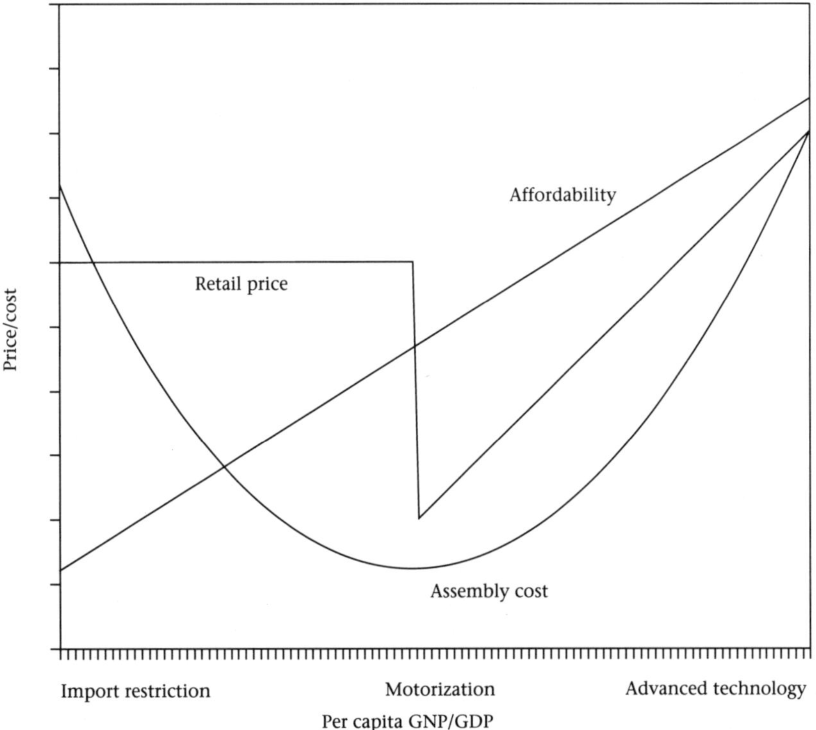

The affordable price does not necessarily coincide with the average retail price in the short-run, however. This was made clear by the analysis in Chapter 8. There is a discretionary range of prices within which manufacturers can decide their model mix. Moreover, the degree of competition in the marketplace ultimately determines the margin for manufacturers and dealers.

The average retail price, therefore, relates not only to the per capita income, but also to the competitive situation in the marketplace. In countries where the competition among manufacturers is fierce, the average retail price reflects this. The average price in the less developed countries is, in general, lower than the price in the advanced markets because the model mix in less developed countries is usually less luxurious and fewer options are required. In this case, the average retail price has a positive correlation with per capita income.

However, in the lesser developed countries, the average retail price of new vehicles can be much higher than in other countries. There are three main reasons for this. First, the bulk of the customers in such markets are the richer citizens who like more option-loaded, stylish vehicles. Second, limited competition among manufacturers and dealers allows creation of a monopolistic margin; since the markets are relatively small, only a few manufacturers and dealers can be sustained. Finally, government policy in such countries tends to impose high tariffs and indirect taxes on imported vehicles.

The general pattern of average prices across countries could be a kinked line, as shown in Figure 18.2, if the production cost curve has the shape drawn in the same figure. Thus, we need to consider the production cost curve. Some assumptions are required if we are to draw this line as it is. First, we must assume that the technology for assembling vehicles can be transferred among countries. Second, in order to have a competitive automotive industry, the countries should have automotive-related industries such as electronics, iron and steel, chemicals, machinery, and design. The final assumption is that all countries can enjoy free trade of vehicles. Based on this assumption, even an automotive company in a small country can achieve full-capacity utilization of its facilities.

The first and third assumptions guarantee that the cost of producing a given model has a positive relationship with the per capita income (which can be considered equivalent to the average wage rate). Of course, the average production cost corresponding to the product mix in each country also shows the same relationship with per capita income. From the second assumption, however, we can see that the least developed countries—those that lack sufficient infrastructure to build vehicles—cannot achieve competitiveness. Even though the assembly cost can be held low because of the technology transfer and a low wage rate, the total production cost could be much higher.

Engines, transmissions, and electronic parts are very difficult to manufacture. Not only manufacturing technology but also management skills play an important role. Without such skills, imports of high-tech components from advanced countries cannot help low-tech companies to utilize those components.

All-in-all, the production cost curve has a "U" shape as drawn in Figure 18.2. The three lines of the figure are highly hypothetical, but they illustrate the dynamics of the global automotive industry very well. The positions of these three lines in Figure 18.2 represent the ideal situation. When the average retail prices are lower than the affordable prices in the relatively developed countries, the vehicles in such markets will be replaced smoothly: that is, over-aged vehicles will be scrapped in a proper cycle.

The continuous production curve suggests that manufacturers in an advanced economy can coexist with manufacturers in less developed countries. Although the latter have a cost advantage over the former, because of the lower wage rate, they specialize in mass production models to meet their domestic demand. The manufacturers in advanced countries tend to make more—and more differentiated—models. Of course, the poorer consumers' demand in the advanced countries may well be met by imports from the less developed countries, and vice versa.

A cost curve lower than the average retail price in the most advanced countries means that manufacturers can earn profits. Although in the most advanced countries, the cost curve is higher than the average retail price, this does not imply that the manufacturers in such countries are losing money—not unless they are monopolizing whole segments. If the cheaper models are supplied by imports while these manufacturers specialize in the high-end products, their profitability could be positive.

When the three lines happen to have the positions shown in Figure 18.2, the global automotive industry can be regarded as very healthy and stable. However, this is not the case with today's situation. If we take the situation of the U.S. industry into consideration, the lines of Figure 18.2 must be modified as shown in Figure 18.3.

In the most advanced countries, the situation shown in Figure 18.3 means that the average retail price is higher than the affordability; this results in an overdependence on used vehicles. Also, the cost curve is intermittent somewhere between the advanced countries and developing countries; this implies that the vehicles manufactured in the developing countries cannot be exported to advanced markets because they cannot meet the high standards set by those advanced countries. An obstacle in the

Figure 18.3 Economic development and the automotive industry: Intermittent technology transfer case

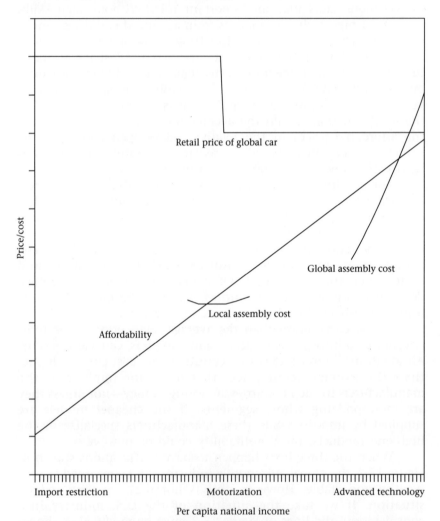

technology transfer thus raises the situation in which consumers' needs can be met only by the manufacturers of the country. But, as the cost curve is higher than the average price line in Figure 18.3, manufacturers cannot earn profits even though the average price is higher than the affordability. This is analogous to the recent situation of the U.S. automotive industry.

The global automotive market is monopolized by the

advanced companies in the case of Figure 18.3. Therefore, they can price their products higher than in the case for Figure 18.2. Their expensive vehicles cannot raise the level of motorization in developing countries as easily as in the case of Figure 18.2, however; they have less opportunity to benefit from such motorization. On the other hand, local manufacturers are specialized in marketing vehicles mainly in their domestic market. Although some of them might have a strong cost-competitiveness in relation to the advanced companies, they cannot take advantage of it. In sum, world welfare is hampered by the lack of technology transfer.

Figure 18.3 suggests several important points. First, the U.S. automotive problem is spreading to other countries. Worldwide, consumers cannot purchase new vehicles at the proper price. The second problem is limited competition among manufacturers, because the companies in the developing countries have difficulty in penetrating into the advanced market. Thus, a greatly expanded role is expected to be played by the advanced manufacturers; the increase in global demand must be satisfied by the expansion of production capacity of the advanced manufacturers. Also, advanced manufacturers need to invest funds in research and development.

Although most of above considerations are not based on quantitative studies, I believe that the situation described by Figure 18.2 is similar to that of the 1970s. In that decade, the Japanese were able to penetrate into the global market through exports. Their technology was not as advanced as that of their American and European rivals, as indicated in Chapter 6, but they could import patents of high-technology components like automatic transmissions, front-wheel drive systems, and electronic fuel-injection systems. Moreover, the cheaper wage rate in Japan, coupled with management efforts like the *kanban* (Just-In-Time) system, helped them to attain cost-competitiveness.

The major difference between the situation shown in Figure 18.2 and that of the 1970s was that the advanced markets had not yet established strict regulatory standards, such as those for emission controls. Rather, the oil crises spotlighted smaller cars because of their fuel efficiency. If the Big Three could have fought the Japanese by taking advantage of their high technology, Japanese penetration might have been limited.

Advanced technology, though, is a double-edged knife. On the one hand, if advanced companies limit its transfer, this hampers global welfare. On the other hand, if advanced manufacturers do not develop new technologies that their rivals in the developing countries cannot easily match, their market share will suffer. However, it is apparent that without advanced technology, automotive companies in the advanced countries cannot coexist with newcomers.[1]

In Chapters 14 and 15 we recommended that the Big Three set up production sites in Mexico in order to manufacture cheap cars. This is an intra-company technology transfer. This strategy fits in with the preceding considerations and will probably work, unless the pricing of cars built in Mexico is improper. Such cars will likely become popular among consumers in developing countries as well.

Therefore, the concept of cars that embody advanced technologies, and which are able to be assembled in the developing countries, should be a focus of the U.S. automotive industry hereafter. This subject is treated in fuller detail in the final chapter.

Note

1. For a treatment of the same situation in a simple economic model, see P. Krugman, "A model of innovation, technology transfer, and the world distribution of income," *Journal of Political Economy*, Vol. 87, No. 2, April 1979, pp. 253–66.

19

Future Cars: A Concept

Both American and global automotive problems are rooted in a common ground. Automotive companies should develop vehicles that can be sold to the less affluent in the U.S. market and that can raise the level of motorization in developing countries. In this chapter, the concept for such a leading vehicle will be discussed.

The necessary features for such vehicles can be summarized as follows:

- The price should be cheap enough for consumers to afford.
- The performance should clear all regulations imposed by the advanced countries.
- The technology utilized to build the model should be transferable to the Newly Industrialized Economies (NIEs).

This car concept is similar to that of the Ford Model T, the Volkswagen Beetle, and the Toyota Corolla; the target of these cars was to widen the spectrum of new car buyers in the American, European, and Japanese markets. But future cars will need to be different from those cars because they will be sold in a variety of countries. This implies that future cars should be able to satisfy a variety of tastes and preferences among consumers. This is a condition that must be met because no single market will be big enough for the manufacturers to achieve the necessary economies of scale. (The Chinese and Russian markets will become big enough in time, but not soon.) The low-end market in the U.S. could be a threshold, but it will continue to be threatened by competition from used vehicles that are large, luxurious, and equipped with high-tech features.

Economies of scale and diversity in the number of models produced are conflicting goals for manufacturers.[1] In pursuing scale merit in production, the number of models needs to be limited as much as possible. To capture consumers with differing needs, however, the number of models should be maximized. How to break through this trade-off is the key to the concept of the future car. Keeping this restrictive condition in mind, a project dubbed the Swatch Cars will be reviewed below.

Swatch is the brand name of watches manufactured by SMH of Switzerland. The innovative idea of Swatch in watchmaking is now seriously being studied worldwide by major automotive companies.[2]

Let us start with a brief history of Swatch in order to understand what kind of innovation the company has achieved.[3] The concept for Swatch was developed when the watchmaking industry in Switzerland was faced with severe competition from Japan and Hong Kong. The global market share of Switzerland in the production of watches and movements was 43 percent in 1970, but fell to 29 percent in 1980 and 13 percent in 1985. The Japanese share in 1985 reached 39 percent, while Hong Kong's reached 22 percent. A dramatic change in technology—from mechanical movement (spring-powered) to quartz (electronic)—caused the decline of the Swiss watchmaking industry. In 1983, SMH was established through the merger of Switzerland's two major watch manufacturers.

SMH embarked on the development of Swatch from early 1980. The aim was to create a new market featuring a "low-price prestige" quartz analog wristwatch. In addition to utilizing the company's original technology in ultra-thin movements, a variety of technologies, such as injection molding and ultrasonic welding, were developed to simplify the production and assembly process. Also, a new marketing strategy was designed to open up a new market: the fashion-oriented younger generation throughout the world. It became apparent that, through massive production, the cost of movements fell remarkably; the higher degree of vertical integration enabled the company to earn sufficient margins.

The key success factor of the Swatch project hinged on how to sell more Swatches in order to secure the benefit of scale economies. The economies of scale stemmed mainly from the mass production of movements, not from the assembly of cases and belts with movements. On the other hand, the styling, color coordination, and design were important to differentiate the individual products. Here lies the innovative feature of Swatch: it was possible for the manufacturer to achieve both economies of scale and product diversity.

From the consumers' point of view, Swatch is not a mere watch. It is, rather, a "fashion accessory that happens to tell the time." A consumer can select the Swatch that best meets his or her tastes and needs. However, the Swatch brand needed to be differentiated so as not to cannibalize other SMH brands, such as Omega, Longines, and Rado—brands that are obviously a major source of SMH's income. In this sense, the different marketing channels and cheap pricing of Swatch were helpful. Moreover, SMH adopted the "model year change-over" policy to motivate periodic replacement of over-aged Swatches, which usually carry a one-year warranty and a battery that should be replaced every three years.

In applying the Swatch concept to the automotive industry, the important issue is whether it is possible to break down the vehicle manufacturing process into two components: "movement" and "other." Fortunately, it is possible to do so. Engines and drivetrains are the major sources of economies of scale in production. Although the styling, color coordination, and upholstery can provide customers with different impressions about a model, diversification with regards to these elements does not totally negate the possibility of exploiting scale economies. Of course, the research and development cost invested in automotive design is far greater than that of Swatch. Therefore, the scale merit in the assembly process is not negligible; the order of magnitude differs significantly from watchmaking.

In a relative sense, however, engines and drivetrains contribute most to the reduction of fixed cost through mass production. As one Japanese expert put it, if the same engine and drivetrain could be used for two consecutive models, that is, eight years, the effect of other cost reduction efforts would be trivial. So far, major Japanese cars have been subjected to a four-year model change cycle, which has made it difficult to recover the development cost. The cumulative production of the same engines and drivetrains for eight years is apparently very desirable.

Thus, application of the Swatch system to automotive production is apparently feasible. The massive production of "movements" will lower the total cost, and the high performance of engines and drivetrains will enable the vehicles to clear governmental regulations in most countries. At the same time, the diversity in design, style, and upholstery can meet varied consumer tastes in many countries. With further improvement for a simpler assembly system, it will be possible to assemble and design Swatch cars in many countries.

However, rapid progress in automotive-making technology as well as frequent changes in governmental regulations may not allow manufacturers to use the same components for such a long period of time. Engines, especially, are a critical factor in determining the performance of vehicles in terms of emission control and fuel efficiency. Swatch cars are likely to be vulnerable to such changes.

This is a major hurdle automotive manufacturers will need to overcome if they are to translate the Swatch concept into a practical application. Some experts think that the Swatch car should be accompanied by a major innovation, similar to the shift from mechanical movement to quartz in watchmaking technology. Electric vehicles are one possible candidate. Of course, the set of battery and motor seem to be a good analogy to the "movement" in watchmaking technology.

Electrically driven vehicles are free from emission control problems, and with improved performance of the battery, they could achieve more efficient energy consumption. Electric vehicles are thus qualified to become a "Swatch car," but it is difficult to say when the necessary breakthrough in battery technology will be completed. The author is of the opinion that we will need to wait another twenty years to develop electric vehicles that offer the same performance as vehicles powered by the current internal-combustion engine.

Other strategists in automotive companies believe that, even with the current gasoline engines, the Swatch concept can be used. They suggest that a Swatch-type commuter or grocery-shopping vehicle can be made. These cars need not have excellent acceleration, long-distance driving capability, or a large storage area. As this type of car can utilize smaller engine displacement, fuel efficiency and emission control can be achieved much more easily than in larger cars. Therefore, changes in governmental regulations might not impede automakers from using Swatch car engines for eight years or longer.

As for safety regulations, smaller cars might find it more difficult to meet a strict standard than larger cars. However, such regulations relate more to the design and style of the car than to the engines and drivetrains used in a particular model. Thus, model change-over could be utilized to strengthen the body from time to time.

The application of the Swatch concept to this type of car seems to be feasible. One potential difficulty that should be pointed out, however, lies not in the technological dimension of the concept but in the managerial dimension. At issue is whether the purchasers of such cars might eventually buy cars of greater added value. As noted earlier, the role of small and low-profit cars for an automotive company is to enlarge the customers' income class. If the customers of such Swatch cars will not upgrade their preference of models as their incomes increase, the total profitability of automotive companies that develop and market Swatch cars cannot be expected to improve.

Moreover, the prevalence of many Swatch cars will result in crowded roads, which could potentially damage the demand for traditional automobiles. In the worst case, large, luxurious cars will be cannibalized by Swatch cars. As cheap and convenient cars will likely become very popular among urban residents, automakers need to come up with a "total" transportation vision for the city before the introduction of such Swatch cars.

At any rate, the Swatch project is a high-risk venture. Under the current demand and supply situation of the global automotive market, however, it has the potential to be a high-return undertaking. Successful founding companies will likely reap enormous profits, and their methodology will become a world standard. Thus, they will be able to take the initiative in this field over the long-term.

Small companies that cannot take the risk will find it tough to enter into the "movement" business. Also, niche-oriented companies and speciality vehicle manufacturers will not have much interest in this business. Possibly a few giant companies in the U.S., EC, and Japan will engage in the "movement" business and form an oligopolistic market.

Although the "movement" manufacturers will be limited to a small number of companies, the profitability will not always be higher than that of other companies. Assemblers who develop an excellent model may well outperform the "movement" suppliers. Even though the gross profitability of "movement" manufacturers is high, they must spend a lot on research and development cost in order to hedge against the risks mentioned above. Thus, with Swatch cars, world automotive manufacturers would be separated into two categories: "movement" manufacturers and "assemblers."

Swatch cars would also generate a more complicated relationship among manufacturers. In order to widen the variety of vehicles with a limited version of "movements," each "movement" manufacturer should purchase "movements" from competitors. This will set up a two-way trade of "movements" among countries: for example, GM might import Toyota's movements at the same time that it exports its own "movements" to Toyota.[4] Also, "movements" will be sold to non-related assemblers in other countries. The brand name of a vehicle would not be related to the make of its "movements." It will be the assemblers who should take the responsibility for the product liability of a model. Thus, some assemblers may form a family for a "movement," such as the VHS family in the video cassette recorder industry.

The introduction of a Swatch system in the assembly line would probably affect the traditional relationships between parts suppliers and automotive companies. Especially in Japan, the *keiretsu* system, in which an automotive company procures parts and components from preferred member suppliers within its group, might become inefficient; the parent "movement" manufacturer cannot control all of the assemblers. It is natural for assemblers to establish a procurement policy based on their own evaluation of suppliers.

One brilliant idea is to recycle some parts of the "movements." Even a quick change in a model will not be harmful to the society. This would also contribute to the replacement of older Swatch cars within a proper period of time.

The Swatch car is just one of the projects under consideration.[5] Much work remains to be done to improve automaking. The notion that automotive technology has already matured should be dismissed immediately. People who are possessed by such an idea are viewing things only from the traditional perspective. The market, the product, and the resources are always evolving. There are new frontiers to be discovered just ahead, if we consider with due deliberation the current problems and inconveniences the industry faces. The company that does not stop challenging an open field is qualified to be the leader of the global automotive industry.

Notes

1. For economic discussions of these features, see K. J. Lancaster, "Socially optimal product differentiation," *American Economic Review*, Vol. 65, No. 4, 1975, pp. 567–85; A. Dixit and J. E. Stiglitz, "Monopolistic competition and optimum product diversity," *American Economic Review*, Vol. 67, No. 3, 1977, pp. 297–308.

2. According to *Automotive News*, Volkswagen planned to develop a small car jointly with SMH. The Swatch car project mentioned here is independent of that project. See "VW quits Swatch car project," *Automotive News*, January 25, 1993, p. 7.
3. C. Pinson, *Swatch*, France: INSEAD-CEDEP, 1987; SMH, *Annual Report*, 1991.
4. For a sophisticated discussion of the international trade of these products, see W. Ethier, "National and international returns to scale in the modern theory of international trade," *American Economic Review*, Vol. 72, No. 3, 1982, pp. 389–405.
5. For a good review of the current activity by global automotive manufacturers, see "The endless road," *The Economist*, October 1992.

Bibliography

Books and Reports

Abernathy, William J., Kim B. Clark, and Alan M. Kantrow, *Industrial Renaissance*, New York: Basic Books, 1983.

Adams, Walter, ed., *The Structure of American Industry*, 7th Edition, New York: Macmillan Publishing, 1982.

Atshuler, Alan, Martin Anderson, Daniel Jones, Daniel Roos, and James Womack, *The Future of the Automobile*, Cambridge, MA: MIT Press, 1984.

Barnett, Steve, ed., *The Nissan Report*, New York: Doubleday, 1992.

Chrysler Corporation, *Annual Report*, various years.

Crandall, Robert W., Howard K. Gruensprecht, Theodore E. Keeler, and Lester B. Lave, *Regulating the Automobile*, Washington, DC: Brookings Institute, 1986.

Economic Report of the President, together with *The Annual Report of the Council of Economic Advisers*, Washington, DC: U.S. Government Printing Office, 1992.

Ford, Henry, *Today and Tomorrow*, Cambridge, MA: Productivity Press, 1986.

Ford Motor Company, *Annual Report*, various years.

General Motors, *Annual Report*, various years.

Hailey, Arthur, *Wheels*, New York: Dell Publishing, 1971.

Halberstam, David, *The Reckoning*, New York: Morrow, 1986.

Iacocca, Lee, *Iacocca: An Autobiography*, New York: Bantam Books, 1984.

Iacocca, Lee, *Lee Iacocca's Talking Straight*, New York: Bantam Books, 1988.

International Monetary Fund, *International Financial Statistics Yearbook*, Washington, DC: IMF, 1991.

Japan Tariff Association, *Japan Exports and Imports, Commodity by Country*, December 1991.

Keller, Maryann, *Rude Awakening*, New York: Harper Perennial, 1989.

Krugman, Paul, *The Age of Diminished Expectations*, Cambridge, MA: MIT Press, 1990.

Lee, Albert, *Call Me Roger*, Chicago: Contemporary Books, 1988.

McAlinder, Sean P., David J. Andrea, Michael S. Flynn, and Brett C. Smith, *The U.S.-Japan Automobile Bilateral 1994 Trade Deficit*, Office for the Study of Automotive Transportation, Transportation Research Institute, University of Michigan, May 1991.

Market Data Book, Detroit: Automotive News, various years.

Moritz, Michael and Barret Seaman, *Going for Broke: The Chrysler Story*, New York: Doubleday, 1981.

Motor Vehicle Manufacturers Association of the United States Inc. (MVMA), *Facts & Figures*, various years.

Nippon Jidousya Kougyoukai (Japan Automobile Manufacturers Association), *Syuyoukoku Jidousya Toukei (Automotive Statistics for Major Countries)*, Japan: Japan Automobile Manufacturers Association, 1992.

Organization for Economic Cooperation and Development (OECD), *The Cost of Restricting Imports, The Automotive Industry*, Paris: OECD, 1987.

Prestowitz, Clyde, Paul Willen, Larry Chimerine, Kevin Kearns, and Robert Cohen, with Mike Flynn and Sean McAlinden, *The Future of the American Auto Industry: It Can Compete. Can It Survive?* Washington, DC: Economic Strategic Institute, 1992.

Rae, John B., *Nissan/Datsun, a History of Nissan Motor Corporation in U.S.A., 1960–80*, New York: McGraw-Hill, 1982.

Roos, Daniel, James P. Womack, and Daniel Jones, *The Machine that Changed the World*, New York: Rawson Associates, 1990.

Sloan Jr., Alfred P., *My Years with General Motors*, New York: Macfaden Books, 1963.

SMH, *Annual Report*, 1991.

Stern, Robert M., Philip H. Trezise, and John Whalley, eds., *Perspective on a U.S.-Canadian Free Trade Agreement*, Washington, DC: Brookings Institute, 1987.

Taub, Eric, *Taurus*, New York: Dutton, 1991.

Thurow, Lester, *Head To Head*, New York: Morrow, 1992.

Toyoda, Eiji, *Toyota—Fifty Years in Motion*, New York: Kodansya International, 1987.

Toyota Motor Corporation, *A History of the First 50 Years*, Japan: Toyota Motor Corporation, 1988.

Toyota Motor Sales, *Toyota, The First Twenty Years in The U.S.A.*, Torrance, CA: Toyota Motor Sales, U.S.A., 1977.

U.S. Department of Commerce, Bureau of the Census, *Annual Survey of Manufactures*, Washington, DC, various years.

U.S. Department of Commerce, Bureau of the Census, *Census of Manufactures*, Washington, DC, 1982 and 1987.

U.S. Department of Commerce, Bureau of the Census, *Statistical Abstract of the United States*, Washington, DC, various issues.

U.S. Department of Labor, Bureau of Labor Statistics, *BLS Handbook of Methods*, Washington, DC, 1988.

U.S. International Trade Commission, *The U.S. Automobile Industry Monthly Report on Selected Economic Indicators*, Washington, DC, various issues.

U.S. International Trade Commission, *U.S. Global Competitiveness: The U.S. Automotive Parts Industry*, Washington, DC, 1987.

Ward's Automotive Yearbook, Detroit: Ward's Communications, various years.

Waring, Stephen P., *Taylorism Transformed*, Chapel Hill: University of North Carolina Press, 1991.

White, Lawrence J., *The Automobile Industry since 1945*, Cambridge, MA: Harvard University Press, 1971.

Articles and Monographs

Automotive News, "Bush gives auto industry break on vapor recovery," March 16, 1992, p. 36.

Automotive News, "VW quits Swatch car project," January 25, 1993, p. 7.

Business Week, "The partners," February 10, 1992, pp. 102–7.

Business Week, "Detroit south," March 16, 1992, pp. 16–21.

Business Week, "The greening of Detroit," April 8, 1992, pp. 54–60.

Business Week, "General Motors: Open all night," June 1, 1992, pp. 66–67.

Business Week, "Chrysler's Neon," May 3, 1993, pp. 116–26.

Carson, Carol S., "GNP: An overview of source data and estimating methods," *Survey of Current Business*, July 1987, pp. 103–26.

Clark, Kim B., W. Bruce Chew, and Takahiro Fujimoto, "Product development in the world auto industry," *Brookings Papers on Economic Activity*, Vol. 3, 1987, pp. 729–81.

Collyns, Charles and Steven Dunaway, "The cost of trade restraints. The case of Japanese automobile exports to the United States," *International Monetary Fund Staff Paper*, Vol. 34, No. 1, 1987, pp. 150–179.

Dixit, A. and J. E. Stiglitz, "Monopolistic competition and optimum product diversity," *American Economic Review*, Vol. 65, No. 3, 1977, pp. 297–308.

The Economist, "The endless road," October 1992.

Ethier, Wilfred J., "National and international returns to scale in the modern theory of international trade," *American Economic Review*, Vol. 72, No. 3, 1982, pp. 389–405.

Feenstra, Robert C., "Voluntary export restraint in U.S. autos, 1980–81: Quality, employment, and welfare effects," in Robert E. Baldwin and Anne O. Krueger, eds., *The Structure and Evolution of Recent U.S. Trade Policy*, Chicago: University of Chicago Press, 1984, pp. 35–65.

Fitzpatrick, Kristin, "Booming Mexico industry awaits decision on free trade," *Ward's Automotive Yearbook 1991*, pp. 112–13.

Fitzwilliams, Jeannette M., "Size distribution of income in 1963," *Survey of Current Business*, April 1964, pp. 3–11.

Forbes, "Platform madness," January 20, 1992, pp. 40–41.

Forbes, "Follow that Ford," April 27, 1992, pp. 44–45.

Homan, Richard, "Popularity contest," *Road & Track*, December 1992, pp. 79–96.

Kennickell, Arthur and Janice Shack-Marquez, "Changes in family finances from 1983 to 1989: Evidence from the survey of consumer finances," *Federal Reserve Bulletin*, January 1992, pp. 1–18.

Krugman, Paul, "A model of innovation, technology transfer, and the world distribution of income," *Journal of Political Economy*, Vol. 87, No. 2, 1979, pp. 253–66.

Krugman, Paul, "The U.S. response to foreign industrial targeting," *Brookings Papers on Economic Activity*, Vol. 1, 1984, pp. 77–131.

Lancaster, Kelvin J., "Socially optimal product differentiation," *American Economic Review*, Vol. 65, No. 4, 1975, pp. 567–85.

McElroy, John, "Ranking the assembly plants," *Automotive Industries*, January 1993, pp. 36–37.

Mannering, Fred and Clifford Winston, "Brand loyalty and the decline of American automobile firms," *Brookings Papers of Economic Activity: Microeconomics*, 1991, pp. 67–114.

Moran, Larry R., "Motor vehicles, model year 1989," *Survey of Current Business*, November 1989, pp. 21–24.

Moran, Larry R., "Motor vehicles, model year 1990," *Survey of Current Business*, November 1990, pp. 27–31.

Moran, Larry R., "Motor vehicles, model year 1991," *Survey of Current Business*, November 1991, pp. 41–45.

Nasar, Sylvia, "Fed gives evidence of 80's gains by richest," *The New York Times*, April 21, 1992.

Parker, Robert P., "A preview of the comprehensive revision of the national income and product accounts: New and redesigned tables," *Survey of Current Business*, October 1991, pp. 20–27.

Pinson, Christian, *Swatch*, France: INSEAD-CEDEP, 1987.

Radner, Daniel B. and John C. Hinrichs, "Size distribution of income in 1964, 1970 and 1971," *Survey of Current Business*, October 1974, pp. 19–31.

Taylor III, Alex, "Can GM remodel itself?" *Fortune*, January 13, 1992, pp. 20–26.

Toyota Motor Corporation, *Nihonsiki Keieini Tuite (On The Japanese Method of Production: The Case of NUMMI)*, a memorandum, 1990.

Ward's Automotive Reports, "Saturn enters 'break-even year'," January 11, 1993, p. 4.

Yoshikawa, Hiroshi, "On the equilibrium yen-dollar rate," *American Economic Review*, Vol. 80, No. 3, 1990, pp. 576–83.

Index

A

A-car, 47–48, 50
Advanced technology strategy, 151–52
Affordable price, 147–51
Age
  average of cars and trucks, 15–16
  distribution of cars, 108–9
American Automobile Manufacturers Association (AAMA). *See* MVMA
Annual median income of families, 71–72, 81
Assemblers, 157–58
Auto loans, 62, 66
Automotive-related industry, 14
Automotive sector, 91
Automotive trade between U.S. and Japan. *See* Trade deficits
Average consumer expenditure, 34, 41–42, 71–72, 88–89. *See also* Price
Average distance traveled per vehicle, 109

B

Brand loyalty, 23–24
Buy American, 119

C

Capacity expansion and wage rate concession, 139
Capacity restraint, 52–53, 76
Capital investment, 51–52
Captive imports, 11
Cars in use, 15, 133, 142
China, 121
Chrysler, 5, 55
Clean Air Act, 110
Compact cars. *See* Large cars
Compensation. *See* Wage rate
Consumer Price Index (CPI), 70, 72
Consumers. *See* New car buyers
Consumer targeting, 87–88, 90–91, 93–98, 117
Coordination failure, 96–97
Cost
  comparison with Japanese, 115–16
  in global dynamics, 148–49
  of satisfying regulatory standards, 26, 112

D

Dealers' markup, 72
Decentralization, 101–2
Demand, 39
  of cars as a percentage of GNP, 34
  in developing countries, 141–42
  forecast model of Japanese company, 63
  forecast for 1990s, 138–39
  global automotive in 1980–92, 140–41
  income elasticity of, 146
  price elasticity of, 60–61
  shift from large cars, 42–43. *See also* Oil crises

Demand and supply in the U.S., 11–12
Depreciation costs, 41–42
Diffusion
rate of cars, 67
rate and per capita income, 144–46
DINK as new car buyers, 64
Domestic orientation, 97
Dumping allegation, 75

E

Earnings. *See* Operating results
East Europe, 121, 145–46
Economy of scale and diversity in taste, 154
Electric vehicles, 156
Emission control, 109–10
Employees, 56
Environment. *See* Emission control
European demand, 140–41

F

Family, 67. *See also* Household
Final sales, 34–35
Financial control, 102–3
Fleet cars, 40–41, 44, 68. *See also* Program cars
Ford, 55
Ford, Henry, 95
Free Trade Agreement between U.S. and Canada, 16
Front-wheel drive cars, 49
Fuel consumption, 108–9
Future cars, 153

G

Gasoline and oil costs, 41
Gasoline prices, 42–43, 44. *See also* Oil crises
General Motors (GM), 3–5, 49–50, 137–38
Gini ratio, 79, 84
Government's role, 112–13
Gulf War, 62

H

Hourly compensation. *See* Wage rate
Household, 67

I

Iaccoca, Lee
on consumer targeting, 87–88
on 1991 operating results, 5
Imports
in automotive statistics, 9
captive, 11
definition of, 17
duty for pickup trucks, 17
of Japanese automobiles, 13. *See also* Penetration of imports
Japan's from Germany, 6
Japan's import duty on cars, 6
ratio, 31–33
Incentives, 60, 63
Income classes, 79–82
of total net worth, 83, 84
Income dispersion, 78, 84, 90
Internal Revenue Service, 75
Inventories, 35. *See also* Statistics: value based

J

Japanese automotive manufacturers
demand and supply, 139–41
future strategy, 125–28

Japanese cars
brands by segment, 44–45
import value, 74–75
market share, 43–44. *See also* Market share
price hikes in 1992 models, 136–37
Job security. *See* White collar recession
Joint investigation in NUMMI, 49, 55
Joint research and promoting competition, 113–14
Junker Car Proposal, 115

K

Keller, Maryann, 18–19
Korean cars, 142
Krugman, Paul, 21

L

Labor productivity, 53–54
difference among the Big Three, 138
international comparison of, 20
in product development, 23
Labor union problems, 21–23, 102, 119. *See also* UAW
Laggard industry. *See* Leading industry
Large cars, 42–43. *See also* Profitability
Leadership, 104
Leading industry, 65–66
Lee, Albert, 19
LH, 136
Light-duty trucks, 134–35
List price, 35, 69
Locally procured components, 114

M

Maintenance costs, 41
Management problems, 18–19
Market recovery, 132
Market share, 13
of Big Three in 1990/91 recession, 61, 135–37
of Japanese cars in upgraded segments, 43–44
segment by segment, 45
Median income. *See* Annual median income
Mexico. *See* NAFTA
Mini van. *See* Light-duty trucks
Monetary income, 84, 98–99
Monopolistic presence, 24. *See also* Technology transfer
Motor Vehicles Manufacturers Association of the United States (MVMA), 17
Movements, 155, 157

N

Neon, 142
New car buyers
demographic analysis for 1991, 59–62
features of, 91–92
shift in the income of, 98
New car sales, 14, 38–39, 46
New United Motors Manufacturing Incorporated (NUMMI), 49–51
Niche strategy, 96
North America Free Trade Agreement (NAFTA), 119–20

O

Oil crises, 42–44
Operating results, 5, 7

Output, 29–31, 34–35

P

Payroll. *See* Wage rate
Penetration of imports, 11–13. *See also* Imports
Personal expenditure, 37–38
Personal spending as a percentage of GDP, 38
Pickup truck. *See* Light-duty trucks
Price
  actual transaction. *See* Average consumer expenditure
  analysis, 69–71
  average retail, 148
  average shipment price index, 135
  of the cheapest model, 73–74
  comparison between domestic and imports, 74–75
  Corolla vs. Corsica, 136
  Japanese export price index, 74–76
  Japanese unit import value, 74–75, 77
  monitoring, 114
  of 1967 comparable car, 71–72
  suggested retail. *See* List price
  weighted average, 70
Pricing in 1990/91 recession, 63–64
Producer Price Index (PPI), 71–72
Production
  domestic, 4, 12
  global vehicle, 9–11
  of Japanese transplants, 11–12
Production capacity, 52–53
Production workers, 53, 56. *See also* Employees
Productivity. *See* Labor productivity
Profitability of compact economy cars, 103
Program cars
  in 1990/91 recession, 65
  in 1992, 133

Q

Quality, 70
Quintile, 80

R

Real family income, 82. *See also* Income classes
Recession, 38, 42, 62–65
Regulatory issues, 24
Retail sales
  analysis of 1992, 131–34
  of the U.S. between 1980 and 1992, 11
Risks
  in car development, 46–47
  in Swatch car project, 156–57
Russia, 121, 145

S

Safety regulation, 110
Saturn, 65, 67–68
Second-hand cars. *See* Used cars
Sloan, Alfred P., Jr., 100
SMH, 154
Statistics
  the nature of automotive, 9–11
  value based, 34–35

Strategic options of the Big Three, 118–19
full-line and international, 120–22
other options, 122–23
Style problems, 23–24
Swatch, 154–55
Swatch car, 155–57

T

Taste of consumers, 36
of the rich, 88–89
Taurus and Sable, 52
Technology problems, 19–21
Technology transfer, 148, 150–51
Tie-ups with foreign manufacturers, 120–21
Trade deficits in 1991, 6, 13–14
Trade-in, 35
Transplant problems, 126–27
Truck. *See* Light-duty trucks

U

Unemployment rate by occupation, 67
United Automobile Workers (UAW), 23, 119. *See also* Labor union problems
Used cars, 15, 35
U.S.-Japan Automotive Summit, 5–7

V

Vehicles, definition of, 30
Voluntary export restraint (VER), 11
quota of Japanese exports, 74–76, 77

W–Z

Wage rate, 21–23, 25, 88–89, 92, 95, 102, 139
international comparison, 22. *See also* Labor union problems
White collar recession, 64–65, 95
World car, 125–26
World regulatory standards, 114

Yen dollar rate, 74–75, 137